ENGINEERING TOOLS, TECHNIQUES AND TABLES

VENTILATION: TYPES, STANDARDS AND PROBLEMS

ENGINEERING TOOLS, TECHNIQUES AND TABLES

Additional books in this series can be found on Nova's website under the Series tab.

Additional E-books in this series can be found on Nova's website under the E-books tab.

MECHANICAL ENGINEERING THEORY AND APPLICATIONS

Additional books in this series can be found on Nova's website under the Series tab.

Additional E-books in this series can be found on Nova's website under the E-books tab.

ENGINEERING TOOLS, TECHNIQUES AND TABLES

VENTILATION: TYPES, STANDARDS AND PROBLEMS

VINCENT A. ROMANO
AND
ALLISON S. DUVAL
EDITORS

Nova Science Publishers, Inc.
New York

For permission to use material from this book please contact us:
Telephone 631-231-7269; Fax 631-231-8175
Web Site: http://www.novapublishers.com

Library of Congress Cataloging-in-Publication Data

Ventilation : types, standards, and problems / editors, Vincent A. Romano and Allison S. Duval.
p. cm.
Includes bibliographical references and index.
ISBN 978-1-61324-424-1 (hardcover : alk. paper) 1. Ventilation. I. Romano, Vincent A. II. Duval, Allison S.
TH7658.V47 2011
697.9'2--dc23

2011013418

Published by Nova Science Publishers, Inc. †New York

CONTENTS

PREFACE

In this book, the authors present topical research in the study of the types, standard and problems of ventilation. Topics discussed include non-conventional energy systems in building design; industrial ventilation systems; pulmonary ventilators as monitoring instruments; influence of the ventilation rate on radon and thoron progeny concentrations in a room; acoustic filters in a power system; RANS and LES turbulence models for indoor airflows and wind towers as natural ventilation devices. (Imprint: Nova)

Chapter 1 - The move towards a de-carbonised world, driven partly by climate science and partly by the business opportunities it offers, will need the promotion of environmentally friendly alternatives, if an acceptable stabilisation level of atmospheric carbon dioxide is to be achieved. This requires the harnessing and use of natural resources that produce no air pollution or greenhouse gases and provides comfortable coexistence of human, livestock, and plants. This study reviews the energy-using technologies based on natural resources, which are available to and applicable in the farming industry. Integral concept for buildings with both excellent indoor environment control and sustainable environmental impacts are reported in the present communication. Techniques considered are hybrid (controlled natural and mechanical) ventilation including night ventilation, thermo-active building mass systems with free cooling in a cooling tower, and air intake via ground heat exchangers. Special emphasis is put on ventilation concepts utilising ambient energy from air ground and other renewable energy sources, and on the interaction with heating and cooling. It has been observed that for both residential and office buildings, the electricity demand of ventilation systems is related to the overall demand of the building and the potential of photovoltaic systems and advanced co-generation units. The focus of the world's attention on environmental issues in recent years has stimulated response in many countries, which have led to a closer examination of energy conservation strategies for conventional fossil fuels. One way of reducing building energy consumption is to design buildings, which are more economical in their use of energy for heating, lighting, cooling, ventilation and hot water supply. Passive measures, particularly natural or hybrid ventilation rather than air-conditioning, can dramatically reduce primary energy consumption. However, exploitation of renewable energy in buildings and agricultural greenhouses can, also, significantly contribute towards reducing dependency on fossil fuels. This chapter describes various designs of low energy buildings. It also, outlines the effect of dense urban building nature on energy consumption, and its contribution to climate change. Measures, which would help to save energy in buildings, are also presented. The massive increases in fuel prices over the last years have however, made any scheme not requiring fuel appear to be more attractive and to be worth reinvestigation. In considering the atmosphere and the oceans as energy sources the four main contenders are wind power, solar energy, biomass energy, wave power, tidal and power from ocean thermal gradients. The renewable energy resources are particularly suited for the provision of rural power supplies and a major advantage is that equipment such as flat plate solar driers, wind machines, etc., can be constructed using local resources and without the advantage results from the feasibility of local maintenance and the general encouragement such local manufacture gives to the build up of small-scale rural based industry. The key factors to reducing and controlling CO_2, which is the major contributor to global warming, are the use of alternative approaches to energy generation and the exploration of how these alternatives are used today and may be used in the future as green energy sources. Even with modest assumptions about the availability of land, comprehensive fuel-wood farming programmes offer significant energy, economic and environmental benefits. These benefits would be dispersed in rural areas where they are greatly needed and can serve as linkages for further rural economic development. Self-renewing resources such as wind, sun, plants and heat from the earth can provide clean abundant energy through the development of renewable technologies. Virtually all regions of the world have renewable resources of one type or another. Research and development investments in the past 25 years in renewable technologies development has lead to important advances in performance and resulting cost effectiveness. Renewable resources currently account for about 9%-10% of the energy consumed in the world; most of this is from hydropower and traditional biomass sources. Wind, solar, biomass and geothermal technologies are cost effective today in an increasing number of markets and are making important steps to broader commercialisation. The present situation is best characterised as one of very rapid growth for wind and solar technologies and of significant promise for biomass and geothermal technologies. Each of the renewable energy technologies is in a different stage of research, development and commer-cialisation and all have differences in current and future expected costs, current industrial base, resource availability and potential impact on energy supply. This chapter discusses the potential for such integrated systems in the stationary and portable power market in response to the critical need for a cleaner energy technology. Anticipated patterns of future energy use and consequent environmental impacts (acid precipitation, ozone depletion and the greenhouse effect or global warming) are comprehensively discussed in this theme. Throughout the theme several issues relating to renewable energies, environment and sustainable development are examined from both current and future perspectives.

Chapter 2 - Ventilation refers to the continuous supply and removal of air with respect to a space. In industry, this is done to control chemical and some physical hazards, as well as to maintain conditions of temperature and relative humidity, which are compatible with human habitation and industrial operation. To move air requires creating a pressure difference between two points. Air will then move from the region of higher to the region of lower pressure, at a rate that depends on the magnitude of the pressure difference and on the impedance to airflow offered by ducts, objects and friction.The pressure difference is created with a fan or blower (mechanical ventilation), or by inducing a density difference through differential heating (natural ventilation). The latter mechanism is the source of wind outdoors but is also used in workplaces where heating sources exist, as in metal refining, melting and casting. The desired performance of the ventilation system will impose a requirements on the magnitude of the pressure difference to be created by the fan or heat source. In many systems, the moving air must capture the airborne contaminant by entraining the surrounding air and conducting it in the desired direction. This leads to requirements for air velocity (speed and direction) and volumetric flow rate. Once captured, the contaminant must be carried away from the inhabited space to a suitable collector or disposal point. This is generally done by forcing the air to flow through ducts where impedance to airflow must be overcome. In some cases, the principal goal of the system is to dilute the contaminant to safe levels by supplying a sufficient volumetric flow rate. Again, there is often a requirement that the direction of air motion be controlled to achieve the desired dilution and worker protection. The design or selection of an industrial ventilation system is strictly related to the laws of fluid flow (fluid-mechanics or fluid-dynamics). In industrial ventilation systems, the involved fluid can be air or a mixture (air and contaminants, which are in solid, liquid or gaseous state). Knowledge of these laws is fundamental for design engineers, and a thorough understanding of basic theory is essential to successful solution of ventilation problem. This paragraph reviews briefly the fundamentals of fluid flow that have significant application to air movement in industrial ventilation systems.

Chapter 3 - The high prevalence of respiratory diseases has led to an increasing demand of assistance services in recent years. This increase also introduced a gradual rise in the admission of patients, what it is particularly important in the case of the respiratory diseases, whose prevalence is high and expected to grow in a near future. Acute lung injury, acute respiratory failure and its most severe form, the acute respiratory distress syndrome, are relatively common in the intensive care unit setting and have a high morbidity and mortality. Chronic obstructive pulmonary disease, for example, affects about 210 million people globally and causes some 3 million deaths annually. These diseases are a major drain on healthcare budgets with 50% of costs accounted for by hospital admissions, much of which could be avoided through development of more responsive models of care allowing earlier recognition and a better treatment. Almost all most critical patients admitted in a Critical Care Unit necessitate ventilatory support. Despite a research shift to non-invasive respiratory support, mechanical ventilation remains an essential tool in the care of the critically ill. The availability of a variety of respiratory modes and somewhat confusing terminology presents a daunting challenge for the physician. But, above all, many of the available treatment modes have not been adequately evaluated and their efficacy appears not completely assessed, by only judging the extent of recent literature reporting experimental evaluations *in vivo* and *in vitro*. In addition, there is paucity of information on the relative merits of those ventilatory strategies that have been studied. From an analysis of recent literature, mechanical, pneumatic, fluid-dynamic and thermo-hygrometric phenomena taking place in the system connecting the patient to the pulmonary ventilator (breathing circuit) seem to influence, and in a not always repeatable way, the pneumatic parameter values actually applied to the patient, the thermo-hygrometric state of the gas mixture, the response time to patient's inspiration effort and the assessment of patient respiratory parameters. Just to give an example, in the last decade, measurement of gas volume (air and oxygen) actually delivered to patient's lungs during mechanical ventilation, especially in case of neonatal ventilation, has received the attention of several scientific authors. The main outcome of this research is that the value of the delivered volume displayed by the ventilator monitoring unit significantly differs from the actual volume that reaches patient's airways. These observations assume increasing importance because of the low mortality associated with ventilation with reduced tidal volumes. It has to be considered that mechanical ventilation is always an interactive process between the machine (pulmonary ventilator, humidifier, breathing circuit) and the patient (characteristics of the respiratory system, kind of patient and pathology, *etc.*), also if the patient is completely passive. Thus, it appears interesting to investigate the above described measurement systems for monitoring and control of mechanical quantities, as it is common practice with other instruments utilized in other fields of science. Moreover, several parameters of patient's pulmonary mechanics are deduced from these measurements. Thus, clinicians figure out quantitative evaluation of the mechanical status of the respiratory system, i.e. respiratory system compliance, airway resistance, work of breathing, from numerical quantities provided by the ventilator monitoring unit and from the observation of the shape of respiratory loops. However, there is no assurance that the air flow and pressure actually delivered to the patient as a function of time correspond to the respiratory pattern set by the physicians. Probably, in order to better understand the reliability and clinical significance of monitoring results, it is necessary to address the study of pulmonary ventilation by considering the system ventilator-circuit-patient as a whole. This chapter, which is intended to represent a food for though for the clinician and researcher dealing with pulmonary ventilation, has the aim to briefly address some causes of possible misleading information which can be obtained from these complex though fascinating machines for vital support.

Chapter 4 - Radon is naturally occurring radioactive element which is member of the ^{238}U radioactive decay series. The main source of radon is surrounding ground, (Eisenbud and Gasell, 1997), from which radon by diffusion and convection reaches surface and enter the atmosphere, (Fleischer, 1997). Building materials which contain higher concentration of radon parent radionuclide – radium, can be intensive sources of radon. ^{222}Rn is the most significant radioisotope because it has been estimated that more than half of the annual effective dose received by human from all sources of ionizing radiation is attributed to ^{222}Rn and its short-lived progeny (NCRP Report 93 1987; ICRP32 1981). It has the half life of 3.825 days and decays emitting alpha particle with energy of 5.49 MeV. When radon enters in dwellings it is usually uniformly distributed in the air, (Urosevic et al, 2008).

Chapter 5 - The acoustic filter is used to filter unwanted frequencies of oscillations from a power system. It is a network with selective transmission for currents from a power system of varying frequency. The unwanted frequencies generated from a power system are removed by using an operational amplifier with different combination of filter arrangements. A new theory for noise protection and security from power systems is presented. A theoretical model for a phenomenon of photovoltaic amplification is formulated. The power systems are classified as per source signals of solar power, electric power, light power, sound power, heat power, fluid power and fire power. The acoustic filters as per source of noise signals are defined. The filters are differentiated as per source signal of unwanted frequencies from solar power, electric power, light power, sound power, heat power, fluid power and fire power. Some results from the simulation model and experiments of glass coated photovoltaic modules are presented. The full scale experiments for a glass coated photovoltaic duct wall section were conducted at Concordia University, Montréal, Québec. The full scale field performance of a photovoltaic duct wall is presented by characterisation of ventilation materials of climate, glass, solar panels, air, wood, polystyrene and electricity. Illustrated examples of noise filtering from power systems are: i) a vocal apparatus; ii) an airflow window with a photovoltaic solar wall; iii) a psychrometric air-conditioner; iv) a telephone line; v) a fire and smoke detection system; vi) sensors and transducers: energy balance for a human brain and vii) indoor environmental quality (IEQ): comfort and health.

Chapter 6 - In buildings low indoor airflows persist, which presents difficulties in modelling of the airflow due to its transitional flow behaviour. The application of a number of turbulence models are investigated in order to evaluate their performance and to provide some issues related to indoor room modelling. The most basic turbulence models are the RANS based models which are quick in providing a solution but are known to fail in transitional flows. Under different conditions such as a channel flow (to predict near wall turbulence), and free jets (to predict diffuser ventilation dispersion), the standard k-ε (SKE) model surprisingly provided sufficient results. Further experimental data was produced which allowed a calibration of the SKE model to further improve its performance. Because the SKE model is so widely used and is known for its generality, the calibrated model was further tested using an enclosed room, and was compared with a Large Eddy Simulation model and experimental data. Finally, the instantaneous results from the LES model highlighted the deficiency of steady simulations using RANS turbulent models which average and smear out the effects of local recirculating eddies.

Chapter 7 - This study investigates the integration of natural ventilation devices into new and existing buildings to meet legislative ventilation requirements. The work focus on the engineering design of a modern wind tower, moreover the individual contribution of each piece of the geometry to the overall effectiveness of the device. Computational Fluid Dynamics (CFD) is used to simulate the air flow and calculate the delivery rates for comparison against occupied space requirements. Presented here for the first time is full scale experimental testing of the device in the natural environment, and the effect of the geometrical variations in an actual building. The experimental work was used to validate the CFD investigations and generate useful performance data as opposed to controlled laboratory conditions testing. The results show that careful component design improves the performance of the device between 27 -45%. Moreover, the wind tower exceeds the minimum legislative requirements at low external wind velocities without the aid of mechanical intervention. Each component is evaluated numerically and experimentally allowing designers and engineers to select and combine to create an individual wind tower to suit the occupied space requirements.

Chapter 8 - Ventilation is the replacement of stale air in a closed space with fresh air. It is a vital requirement for the comfort and health of closed space occupants. The reasons why ventilation is so important are to supply oxygen for breathing (respiration), to dilute pollutants such as body odours, to exhaled carbon dioxide, to remove unwanted heat from the closed space, to supply air for combustion appliances, to lower the relative humidity and so avoid condensation and to clear smoke in the event of fire. Ventilation systems may be classified according their ability to supply and withdraw air from the closed space ventilated. It is common to different between the short cut principle, mixed principle, displacement principle and piston principle. Also, there are essentially some ways of ventilating any closed space as ventilation by displacement, ventilation by diffusion, natural ventilation, mechanical or forced ventilation, extract ventilation, supply ventilation and balanced ventilation.

Chapter 9 - Indoor air is a fundamental parameter to be considered at the building design moment. In this sense, an adequate selection of a ventilation method must be done among natural, mechanical and hybrid ventilations. In recent research works, based on the tracer gas decay method, these three ventilation methodologies were compared in urban environments to define their performances in accordance with the nominal time constant, which is the inverse value of air-exchange rates and air-exchange efficiency. These results showed that hybrid ventilation is associated with rather higher nominal time constants and higher air exchange efficiency values than natural cross-ventilation. It was obtained that ventilation efficiencies seem to be higher when the nominal time constants are higher or the air-exchange rates are lower. In this chapter some different methods of natural ventilation and what the current research works state, step by step, are shown. In particular, these steps are classified on the basis of field measurements, laboratory experimental data and simulation tools. Finally, future ventilation research topics like energy saving, thermal comfort and personalized ventilation methods are presented. The general obtained conclusion is that residential mixing ventilation and industrial hybrid displacement ventilation are very interesting alternatives. Furthermore, this discusses how a good agreement between developed models and experimental results is obtained and, subsequently, it can be employed to predict the relative performances of ventilation systems.

In: Ventilation: Types, Standards and Problems
Editors: V. A. Romano and A. S. Duval, pp. 1-48
ISBN 978-1-61324-424-1

Chapter 1

NON-CONVENTIONAL ENERGY SYSTEMS IN BUILDINGS DESIGNS: HEATING, VENTILATION, AIR-CONDITIONING AND ENVIRONMENTAL POLLUTION CONTROL

Abdeen Mustafa Omer
Forest Road West, Nottingham,
United Kingdom

ABSTRACT

The move towards a de-carbonised world, driven partly by climate science and partly by the business opportunities it offers, will need the promotion of environmentally friendly alternatives, if an acceptable stabilisation level of atmospheric carbon dioxide is to be achieved. This requires the harnessing and use of natural resources that produce no air pollution or greenhouse gases and provides comfortable coexistence of human, livestock, and plants. This study reviews the energy-using technologies based on natural resources, which are available to and applicable in the farming industry. Integral concept for buildings with both excellent indoor environment control and sustainable environmental impacts are reported in the present communication. Techniques considered are hybrid (controlled natural and mechanical) ventilation including night ventilation, thermo-active building mass systems with free cooling in a cooling tower, and air intake via ground heat exchangers. Special emphasis is put on ventilation concepts utilising ambient energy from air ground and other renewable energy sources, and on the interaction with heating and cooling. It has been observed that for both residential and office buildings, the electricity demand of ventilation systems is related to the overall demand of the building and the potential of photovoltaic systems and advanced co-generation units. The focus of the world's attention on environmental issues in recent years has stimulated response in many countries, which have led to a closer examination of energy conservation strategies for conventional fossil fuels. One way of reducing building energy consumption is to design buildings, which are more economical in their use of energy for heating, lighting, cooling, ventilation and hot water supply. Passive measures, particularly natural or hybrid ventilation rather than air-conditioning, can dramatically reduce primary energy consumption. However, exploitation of renewable energy in buildings and agricultural greenhouses can, also, significantly contribute towards reducing dependency on fossil fuels. This chapter describes various designs of low energy buildings. It also, outlines the effect of dense urban building nature on energy consumption, and its contribution to climate change. Measures, which would help to save energy in buildings, are also presented. The massive increases in fuel prices over the last years have however, made any scheme not requiring fuel appear to be more attractive and to be worth reinvestigation. In considering the atmosphere and the oceans as energy sources the four main contenders are wind power, solar energy, biomass energy, wave power, tidal and power from ocean thermal gradients. The renewable energy resources are particularly suited for the provision of rural power supplies and a major advantage is that equipment such as flat plate solar driers, wind machines, etc., can be constructed using local resources and without the advantage results from the feasibility of local maintenance and the general encouragement such local manufacture gives to the build up of small-scale rural based industry. The key factors to reducing and controlling CO_2, which is the major contributor to global warming, are the use of alternative approaches to energy generation and the exploration of how these alternatives are used today and may be used in the future as green energy sources. Even with modest assumptions about the availability of land, comprehensive fuel-wood farming programmes offer significant energy, economic and environmental benefits. These benefits would be dispersed in rural areas where they are greatly needed and can serve as linkages for further rural economic development. Self-renewing resources such as wind, sun, plants and heat from the earth can provide clean abundant energy through the development of renewable technologies. Virtually all regions of the world have renewable resources of one type or another. Research and development investments in the past 25 years in renewable technologies development has lead to important advances in performance and resulting cost effectiveness. Renewable resources currently account for about 9%-10% of the energy consumed in the world; most of this is from hydropower and traditional biomass sources. Wind, solar, biomass and geothermal technologies are cost effective today in an increasing number of markets and are making important steps to broader commercialisation. The present situation is best characterised as one of very rapid growth for wind and solar technologies and of significant promise for biomass and geothermal technologies. Each of the renewable energy technologies is in a different stage of research, development and commer-cialisation and all have differences in current and future expected costs, current industrial base, resource availability and potential impact on energy supply. This chapter discusses the potential for such integrated systems in the stationary and portable power market in response to the critical need for a cleaner energy technology. Anticipated patterns of future energy use and consequent environmental impacts (acid precipitation, ozone depletion and the greenhouse effect or global warming) are comprehensively discussed in this theme. Throughout the theme several issues relating to renewable energies, environment and sustainable development are examined from both current and future perspectives.

Keywords: Energy, renewable energy technologies, energy efficiency comfort, Built environment, ventilation, sustainable development, sustainable environmental impact, global warming, emissions.

1. INTRODUCTION

Globally, buildings are responsible for approximately 40% of the total world annual energy consumption [1]. Most of this energy is for the provision of lighting, heating, cooling, and air conditioning. Increasing awareness of the environmental impact of CO_2 and NO_x emissions and CFCs triggered a renewed interest in environmentally friendly cooling, and heating technologies. Under the 1997 Montreal Protocol, governments agreed to phase out chemicals used as refrigerants that have the potential to destroy stratospheric ozone. It was therefore considered desirable to reduce energy consumption and decrease the rate of depletion of world energy reserves and pollution of the environment. This chapter discusses a comprehensive review of energy sources, environment and sustainable development. This includes all the renewable energy technologies, energy efficiency systems, energy conservation scenarios, energy savings and other mitigation measures necessary to reduce climate change.

One way of reducing building energy consumption is to design buildings, which are more economical in their use of energy for heating, lighting, cooling, ventilation and hot water supply. Passive measures, particularly natural or hybrid ventilation rather than air-conditioning, can dramatically reduce primary energy consumption [2]. However, exploitation of renewable energy in buildings and agricultural greenhouses can, also, significantly contribute towards reducing dependency on fossil fuels. Therefore, promoting innovative renewable applications and reinforcing the renewable energy market will contribute to preservation of the ecosystem by reducing emissions at local and global levels. This will also contribute to the amelioration of environmental conditions by replacing conventional fuels with renewable energies that produce no air pollution or greenhouse gases.

The provision of good indoor environmental quality while achieving energy and cost efficient operation of the heating, ventilating and air-conditioning (HVAC) plants in buildings represents a multi variant problem. The comfort of building occupants is dependent on many environmental parameters including air speed, temperature, relative humidity and quality in addition to lighting and noise. The overall objective is to provide a high level of building performance (BP), which can be defined as indoor environmental quality (IEQ), energy efficiency (EE) and cost efficiency (CE):

- Indoor environmental quality is the perceived condition of comfort that building occupants experience due to the physical and psychological conditions to which they are exposed by their surroundings. The main physical parameters affecting IEQ are air speed, temperature, relative humidity and quality.
- Energy efficiency is related to the provision of the desired environmental conditions while consuming the minimal quantity of energy.
- Cost efficiency is the financial expenditure on energy relative to the level of environmental comfort and productivity that the building occupants attained. The overall cost efficiency can be improved by improving the indoor environmental quality and the energy efficiency of a building.

An approach is needed to integrate renewable energies in a way to meet high building performance. However, because renewable energy sources are stochastic and geographically diffuse, their ability to match demand is determined by adoption of one of the following two approaches [2]: the utilisation of a capture area greater than that occupied by the community to be supplied, or the reduction of the community's energy demands to a level commensurate with the locally available renewable resources. For a northern European climate, which is characterised by an average annual solar irradiance of 150 Wm^{-2}, the mean power production from a photovoltaic component of 13% conversion efficiency is approximately 20 Wm^{-2}. For an average wind speed of 5 ms^{-1}, the power produced by a micro wind turbine will be of a similar order of magnitude, though with a different profile shape. In the UK, for example, a typical office building will have a demand in the order of 300 $kWhm^{-2}yr^{-1}$. This translates into approximately 50 Wm^{-2} of façade, which is twice as much as the available renewable energies [3]. Thus, the aim is to utilise energy efficiency measures in order to reduce the overall energy consumption and adjust the demand profiles to be met by renewable energies. For instance, this approach can be applied to greenhouses, which use solar energy to provide indoor environmental quality. The greenhouse effect is one result of the differing properties of heat radiation when it is generated at different temperatures. Objects inside the greenhouse, or any other building, such as plants, re-radiate the heat or absorb it. Because the objects inside the greenhouse are at a lower temperature than the sun, the re-radiated heat is of longer wavelengths, and cannot penetrate the glass. This re-radiated heat is trapped and causes the temperature inside the greenhouse to rise. Note that the atmosphere surrounding the earth, also, behaves as a large greenhouse around the world. Changes to the gases in the atmosphere, such as increased carbon dioxide content from the burning of fossil fuels, can act like a layer of glass and reduce the quantity of heat that the planet earth would otherwise radiate back into space. This particular greenhouse effect, therefore, contributes to global warming. The application of greenhouses for plants growth can be considered one of the measures in the success of solving this problem. Maximising the efficiency gained from a greenhouse can be achieved using various approaches, employing different techniques that could be applied at the design, construction and operational stages. The development of greenhouses could be a solution to farming industry and food security.

Energy has been a vital input into the economic and social development. However, one third of the world population, living in developing and threshold countries, has no access to electricity. These people mostly live in remote and rural areas with low population density, lacking even the basic infrastructure. Accordingly, utility grid extension is not a cost-effective option and sometimes technically not feasible. Therefore, it is imperative to look for sustainable (i.e., cost-effective, environmentally benign and reliable) sources of energy for the development of these regions. Using locally available renewable energy sources (especially solar irradiation that is characterised by a sufficient availability on a daily basis), which are of high potential in most of these regions, offers a strategic solution for their techno-economic development. From the point of view of technology, the design of system technology that meets electrification requirements and fulfils, if necessary, the requirements of integration into alternative current (AC) supply grids, has to be considered.

The modernisation of the system components and their power ranges which allow easy expandability of the supply structure, the standardisation of interfaces and the hybridisation by integration of different energy converters in order to increase the power availability, represent the most important measures from the point of view of system technology. Moreover, the use of renewable energy sources is essentially made easier if the existing reliable AC- technical standards of construction and extension of conventional electricity supply systems are adopted. Therefore, incompatibility cannot be taken as a reason to reduce the dissemination of renewable systems.

In the early years of photovoltaic (PV) history stand-alone as well as grid-connected PV-systems had been built as individual items or unique masterpieces. Until recent years the realisation of PV-systems still was characterised by monolithic system concepts resulting in a costly design and engineering process. Consequently, a large number of different PV system components (e.g., inverter) were developed each tailored for the use in the dedicated application with its specific parameters (e.g., input voltage/current).

The exploitation of the energetic potential (solar and wind) for the production of electricity proves to be an adequate solution in isolated regions where the extension of the grid network would be a financial constraint. The use of wind as alternative energy source is increasing and research and development about this clean and unlimited resource is being carried out on various levels.

Likewise, energy savings from the avoidance of air conditioning can be very substantial. Whilst day-lighting strategies need to be integrated with artificial lighting systems in order to become beneficial in terms of energy use, reductions in overall energy consumption levels by employment of a sustained programme of energy consumption strategies and measures would have considerable benefits within the buildings sector. The perception often given however is that rigorous energy conservation as an end in itself imposes a style on building design resulting in a restricted aesthetic solution. It would perhaps be better to support a climate sensitive design approach, which encompassed some elements of the pure conservation strategy together with strategies, which work with the local ambient conditions making use of energy technology systems, such as solar energy, where feasible. In practice, low energy environments are achieved through a combination of measures that include:

- The application of environmental regulations and policy.
- The application of environmental science and best practice.
- Mathematical modelling and simulation.
- Environmental design and engineering.
- Construction and commissioning.
- Management and modifications of environments in use.

In summary, achieving low energy building requires comprehensive strategy that covers; not only building designs, but also considers the environment around them in an integral manner. Major elements for implementing such a strategy are as follows:

1.1. Efficiency Use of Energy

- Climate responsiveness of buildings.
- Good urban planning and architectural design.
- Good house keeping and design practices.
- Passive design and natural ventilation.
- Use landscape as a means of thermal control.
- Energy efficiency lighting.

- Energy efficiency air conditioning.
- Energy efficiency household and office appliances.
- Heat pumps and energy recovery equipment.
- Combined cooling systems.
- Fuel cells development.

1.2. Utilise Renewable Energy

- Photovoltaics.
- Wind energy.
- Small hydros.
- Waste-to-energy.
- Landfill gas.
- Biomass energy.
- Biofuels.

1.3. Reduce Transport Energy

- Reduce the need to travel.
- Reduce the level of car reliance.
- Promote walking and cycling.
- Use efficient public mass transport.
- Alternative sources of energy and fuels.

1.4. Increase Awareness

- Promote awareness and education.
- Encourage good practices and environmentally sound technologies.
- Overcome institutional and economic barriers.
- Stimulate energy efficiency and renewable energy markets.

2. Energy from Waste

Measures to maximise the use of high-efficiency generation plants and on-site renewable energy resources are important for raising the overall level of energy efficiency. The world's view of waste has changed dramatically in recent years and it is now seen as a source to feed the ever-growing demand for energy (Figure 1). The road from the initial concept to the production of the first kilowatt of power is long and has many challenges, not least the need for adequate funding. Scientific research evidence, public awareness and increased levels of participation in environmental campaigning have led to governments' worldwide implementing regulations and legislation. Examples include:

- EU landfill diversion directive.
- Recycling targets.
- Climate change regulations.

The demand for nuclear power generation, wind farms, solar power and so on is now unstoppable and has created a whole new market, though each has its own challenges (Figure 2). The waste collection, transfer and landfill disposal business comprise a mature, slow-growth industry. Economic drivers to developing the waste and renewable energy sector have included:

- Waste disposal and landfill gate fees/landfill tax.
- Penalties/avoidance schemes (e.g., landfill allowance schemes and fines, carbon trading).
- Energy prices.
- Investments subsidies.

When considering the demand and opportunity in today's marketplace these points are prevalent:

- The demand for renewable energy is not going to go away.
- The public feeling is that governments across the world are responsible.
- The pressure caused by diminishing fossil fuel supplies is increasing.
- Investment funds are increasingly available from traditional sources.
- The needs for new technologies that can deliver carbon reduction and waste reduction outcomes are increasingly bankable, which opens up the market for all.

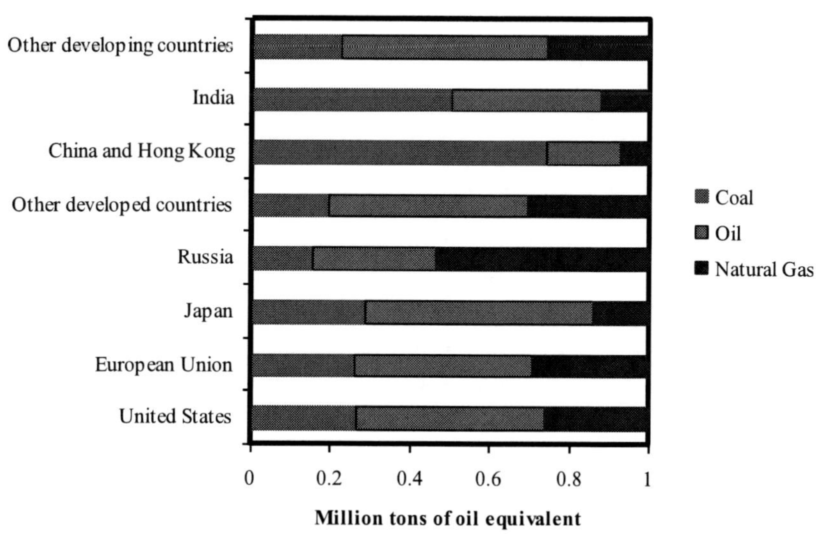

Figure 1. Global fossil fuel consumption.

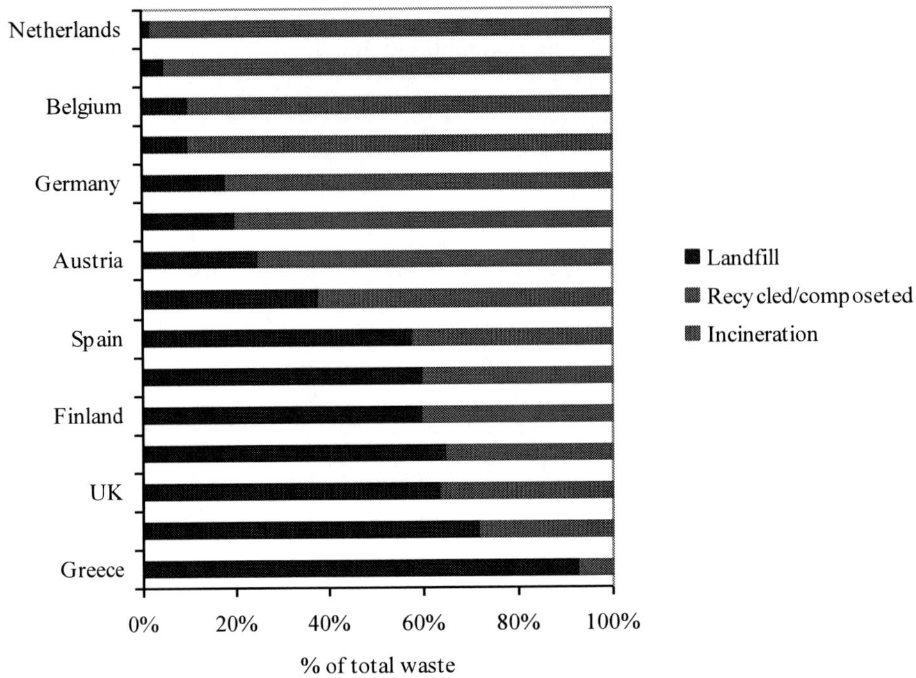

Figure 2. Municipal waste management in the European Union.

Financial institutions across most global markets are gearing themselves up for the environmental revolution. Within the waste to renewable energy sector, history has shown a hesitancy to invest in projects not supported by four things:

- Adequate independent technology due diligence.
- Security of waste input and power off-take contracts.
- A site with planning permission.
- A reference plant, preferably at scale.

Reviewing the evolution of municipal sewage waste (MSW) management in general, waste collection has tended to progress from incomplete collection through to complete collection and finally to collection with separation into different waste streams. In turn, waste treatment has progressed from ad-hoc decentralised disposal to a strategy more dependent on controlled treatment and disposal, including the use of sanitary landfilling accompanied by waste reduction strategies. In developed countries, this evolution has taken place over a period of about 30-40 years. The standard of liner design and construction standard are summarised in Table 1.

One of the negative results of growing prosperity worldwide has been an increase in waste generation from year to year. In response, policy-makers and researchers are examining how best to decouple waste growth and economic growth. In both developed and developing countries sanitary landfill sites can be operated in such a way that danger to residents and the environment, from leachate, odours, fire and explosion is almost entirely eliminated. Table 2 summarised different parameters in waste compaction. Waste professionals use cross wrap machinery for its reliability and efficiency in storage and transport of waste materials.

Table 1. Comparison of basic requirements for bottom liners in MSW sanitary landfills

Liner system requirement	Leachate drainage layer	Geomembrane line	Compacted clay layer
US EPA Standard (40CFR258)	K >1x10^{-4} ms^{-3} Thickness 0.3 m	Thickness ≥0.75 mm Recommended 1.5 mm HDPE	K ≤ lx 10^{-6} ms^{-3} Thickness ≈60 cm
EU Landfill Directive (1999/31/Dec)	Thickness 0.5 m	Not specified. Yet liner thickness should be 100 cm K≤ lx 10^{-6} ms^{-3}	With HDPE liner, thickness of clay layer > 50 cm
German Standard (TASI 1993)	K >1x10^{-3} ms^{-3} Thickness 0.3 m	Thickness ≥2.5 mm HDPE	K ≤ 5x 10^{-10} ms^{-3} Thickness 3x25 cm
Chinese Standard (CJJ 113-2007)	K >1x10^{-3} ms^{-3} Thickness >0.3 m	Thickness ≥1.5 mm HDPE	K ≤ lx 10^{-6} ms^{-3} Thickness 75 cm

Table 2. The different parameters in waste compaction

Refuse	Item size Organic components Inert substances Slurry
Application technique	Thin layer operation Face operation Pushing distance
Compaction machine	Compactor Operating weight Wheel design
Types of waste disposal site	Pit type waste disposal site Raised refused disposal site Height of the refuse disposal site Refuse load
Weather conditions	Precipitation Temperature

2.1. Waste Shredding

With the demand for faster and more efficient recycling technologies showing no signs of abating, the market for faster, more efficient shredding equipment is of course on the up. To the man on the street the term shredding most likely brings to mind the transformation of business documents, bank and credit card statements into a bird's nest of paper- a practice now relies on worldwide to prevent fraudsters accessing the personal financial data and sensitive information. It means big business, as shredding of waste is common practice across almost all areas of the waste industry. Far from focusing simply on paper, shredding is a disposal technique for everything from agriculture to household waste and electrical to industrial waste.

The overall trend in today's market tends to be 'shred first and sort later'. Shredding of waste material as a precursor to sorting is useful for two reasons. It reduces the size of the waste, allowing for greater ease of transportation, but perhaps more importantly – at a time when recycling as much materials as effectively as possible is paramount- it allows for more effective sorting afterwards. And logically, effective sorting equals greater opportunity for recycling (Figure 3).

While every customer on the lookout for a shredder is interested in efficiency, one thing that will also attract a potential buyer is energy efficiency. With the large environmental challenges facing the world today, waste industry professionals are increasingly aware of the need to make sure their business are as kind to the environment as possible. The operation of large-scale equipment such as shredding machines naturally uses a large amount of power and machines, which can run effectively on a lesser amount, have an advantage over their competitors.

Even with modest assumptions about the availability of land, comprehensive fuel-wood farming programmes offer significant energy, economic and environmental benefits. These benefits would be dispersed in rural areas where they are greatly needed and can serve as linkages for further rural economic development. The nations, as a whole would benefit from savings in foreign exchange, improved energy security, and socio-economic improvements. With a nine-fold increase in forest – plantation cover, the nation's resource base would be greatly improved. The international community would benefit from pollution reduction, climate mitigation, and the increased trading opportunities that arise from new income sources. The aim of any modern biomass energy systems must be:

- To maximise yields with minimum inputs.
- Utilisation and selection of adequate plant materials and processes.
- Optimum use of land, water, and fertiliser.
- Create an adequate infrastructure and strong R and D base.

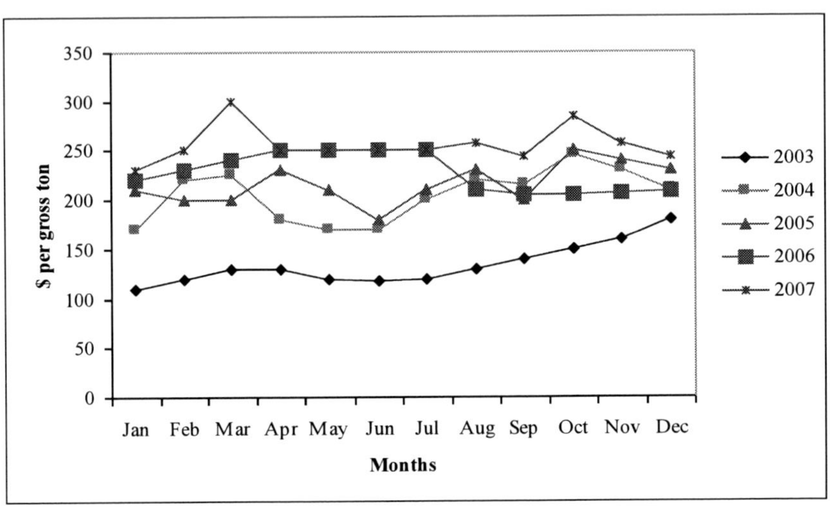

Figure 3. Monthly averages paper scraps.

2.2. Biomass CHP

Combined heat and power (CHP) installations are quite common in greenhouses, which grow high-energy, input crops (e.g., salad vegetables, pot plants, etc.). Scientific assumptions for a short-term energy strategy suggest, that the most economically efficient way to replace the thermal plants is to modernise existing power plants to increase their energy efficiency and to improve their environmental performance. However, utilisation of wind power and the conversion of gas-fired CHP plants to biomass would significantly reduce the dependence on imported fossil fuels. Although a lack of generating capacity is forecasted in the long-term, utilisation of the existing renewable energy potential and the huge possibilities for increasing energy efficiency are sufficient to meet future energy demands in the short-term.

A total shift towards a sustainable energy system is a complex and long process, but is one that can be achieved within a period of about 20 years. Implementation will require initial investment, long-term national strategies and action plans. However, the changes will have a number of benefits including a more stable energy supply than at present and major improvement in the environmental performance of the energy sector, and certain social benefits. A vision used a methodology and calculations based on computer modelling that utilised:

- Data from existing governmental programmes.
- Potential renewable energy sources and energy efficiency improvements.
- Assumptions for future economy growth.
- Information from studies and surveys on the recent situation in the energy sector.

In addition to realising the economic potential identified by the National Energy Savings Programme, a long-term effort leading to a 3% reduction in specific electricity demand per year after 2020 is proposed. This will require further improvements in building codes, and continued information on energy efficiency.

The environmental Non Governmental Organisations (NGOs) are urging the governments to adopt sustainable development of the energy sector by:

- Diversifying of primary energy sources to increase the contribution of renewable and local energy resources in the total energy balance.
- Implementing measures for energy efficiency increase at the demand side and in the energy transformation sector.

Methane is a primary constituent of landfill gas (LFG) and a potent greenhouse gas (GHG) when released into the atmosphere. Globally, landfills are the third largest anthropogenic emission source, accounting for about 13% of methane emissions or over 818 million tones of carbon dioxide equivalent ($MMTCO_2e$) [4] as shown in Figure 4.

The price of natural gas is set by a number of market and regulatory factors that include:

Supply and demand balance and market fundamentals, weather, pipeline availability and deliverability, storage inventory, new supply sources, prices of other energy alternatives and regulatory issues and uncertainty.

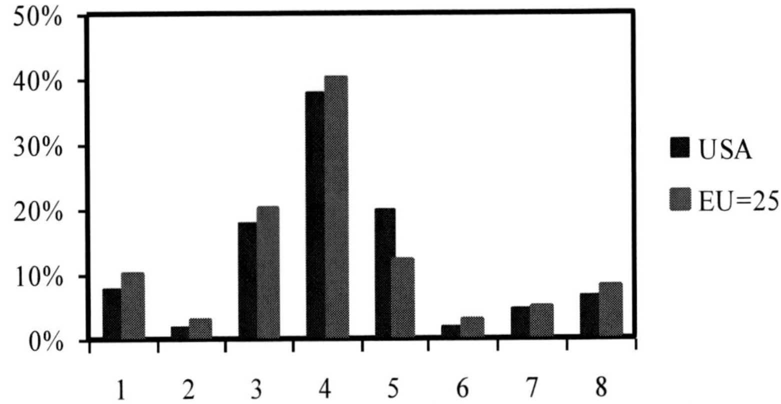

1 Food, 2 Textile, 3 Pulp and paper, 4 Chemicals, 5 Refining, 6 Minerals, 7 Primary metals, and 8 Others.

Figure 4. Distribution of industrial CHP capacity in the EU and USA.

Classic management approaches to risk are well documented and used in many industries. This includes the following four broad approaches to risk:

- Avoidance includes not performing an activity that could carry risk. Avoidance may seem the answer to all risks, but avoiding risks also means losing out on potential gain.
- Mitigation/reduction involves methods that reduce the severity of potential loss.
- Retention/acceptance involves accepting the loss when it occurs. Risk retention is a viable strategy for small risks. All risks that are not avoided or transferred are retained by default.
- Transfer means causing another party to accept the risk, typically by contract.

3. Wind Energy

Eventually renewable energies will dominate the world's energy supply system. There is no real alternative. Mankind cannot indefinitely continue to base its life on the consumption of finite energy resources. There are numerous factors that influence the overall prospects for the wind industry, though in the end it is the economics that will be the deciding factor (Table 3). The most important issues identified:

- Assessment of previous patterns of market development in similar markets.
- Increased engagement of utilities and large energy companies.
- National energy plans and government support for renewable energy.
- Technical development.
- Growth in market and the present dynamics of the industry.
- Information about specific large projects.
- Assessment of wind resources and how they can be used.

Table 3. Market shares 2005-2007

Years		2005		2006		2007	
Manufacturer	Country	Supplied	Share	Supplied	Share	Supplied	Share
Vestas	Denmark	3186	27.6%	4239	28.2%	4503	22.8%
Ge Wind	US	2025	17.5%	2326	15.5%	3283	16.6%
Gamesa	Spain	1474	12.8%	2346	15.6%	3047	15.4%
Enercon	Germany	1640	14.2%	2316	15.4%	2769	14.0%
Suzton	India	700	6.1%	1157	7.7%	2082	10.5%
Siemens	Denmark	629	5.4%	1103	7.3%	1397	7.1%
Acciona	Spain	224	1.9%	426	2.8%	873	4.4%
Goldwind	China	132	1.1%	416	2.8%	830	4.2%
Nordex	Germany	298	2.6%	505	3.4%	676	3.4%
Sinovel	China	3	0.0%	75	0.5%	671	3.4%
Others		1032	8.9%	1094	7.3%	2076	10.5%
Total		11343	98%	16003	107%	22207	112%

Most of these factors are favourable for the industry at the moment. There is strong political support for wind energy, both as engineering and supply chain problems that have been associated with rapid growth in the past.

Economic projections are difficult at the best of times, when economies are relatively stable and a reference 'business as usual' case can be used. However, there are numerous signals that the world faces very turbulent economic conditions for a while- a credit crunch may make some project finance difficult and the shortage of raw materials could lead to supply chain difficulties. However, the rapidly escalating price of oil is focusing a lot of attention on the price of energy and the hedge of electricity supply without a fuel cost is likely to become increasingly attractive to many companies and utilities. At some stage, rising fuel costs could lead to demand for wind energy becoming almost infinite. The main factors expected to influence the continuing growth of the wind sector are:

- The economies of the transition states (Russia and Central Asia) will start to grow.
- Increasing energy demand in Asia and South America.
- Oil prices will continue to remain high as will demand for fossil fuels.
- Continuing competitiveness of wind with fossil fuels.
- Many countries may find they are well off their international CO_2 reduction commitments and need to install some new renewable capacity very quickly.
- Security of supply questions will continue to support wind power.
- Deregulated markets will remove excess conventional power capacity and new capacity is likely to be more expensive than wind.

While wind energy can still seem a small industry compared with conventional power generation, the achievement of 1% of world electricity generation is potentially significant. In individual markets such as Denmark, Germany and Spain reaching 1% has been a breakthrough figure, establishing a critical mass and being followed by further rapid growth in each year market. If the same pattern is seen with world wind energy demand and the industry continues to establish itself as a significant player in the energy sector and pushes on rapidly to 30% of world electricity demand and beyond, then the glass should be seen as half full.

Wind energy is one of the low investments high yielding sources of power generation. The future of wind energy is extremely bright and there is no doubt that in the renewable energy sector, wind power would play a predominant role in adding to the national grids clean and non-polluting energy in the coming years (Table 4).

Table 4. Installed capacity per year

Year	Europe (MW)	World (MW)
Before 2000	9.413	13.954
2000	13.306	18.449
2001	17.812	24.927
2002	23.832	32.037
2003	29.301	40.301
2004	34.725	47.912
2005	40.897	59.320
2006	48.628	74.517
2007	57.136	94.593
2008	66.785	120.458
2009	78.514	151.753
2010	93.590	191.318

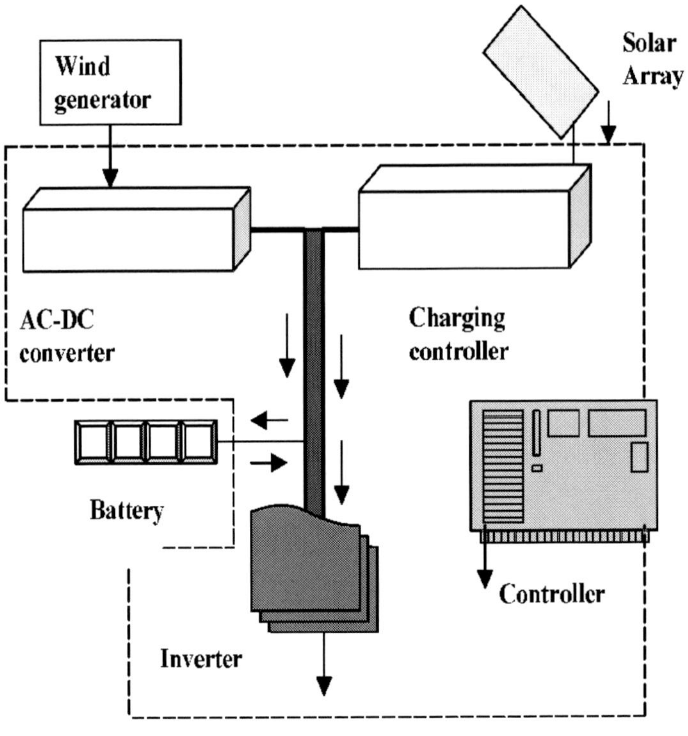

Figure 5. Wind-Photovoltaic hybrid generation systems.

In recent years, demand for the micro wind turbines, of the output below 1 kW, is on the increase as monuments and educational materials. Most of the micro wind turbine that has a diameter under 1.0 m is low blade tip speed ratio type on the market, by the problem of the frequency, the safety and the blade noise. In these circumstances, it would be necessary to develop the system characteristics of micro wind turbines for the purpose of much higher performance in spite of the low Reynolds number regions. Wind power generation is characterised by its stochastic nature, whereby supply and demand, in small grid systems in particular, mostly do not match. The combination of wind power with a second complementary power generation and/or direct/indirect storage technology therefore has, in principle, considerable potential. Wind-diesel, wind-water desalination and wind power in combination with hydrogen production are all potential options that have been high on the international renewable energy agenda for several years. A small-scale wind-PV hybrid power generator system for dairy farm is shown in Figure 5, to verify the possibilities to apply a power generating system and heating source for dairy farm. It is possible to apply the system for power supply and heat source to melt snow and process fertiliser.

Global investment in renewables and energy efficiency now outpaces that for nuclear energy. Renewables also accounted for more than a fifth of new generation capacity built in 2007. The renewable energy and energy efficiency sectors seeing a level of commercial investment that most thought unattainable just a few years back. The risk factors pointing towards a 'bust' as identified and summed up in Figure 6.

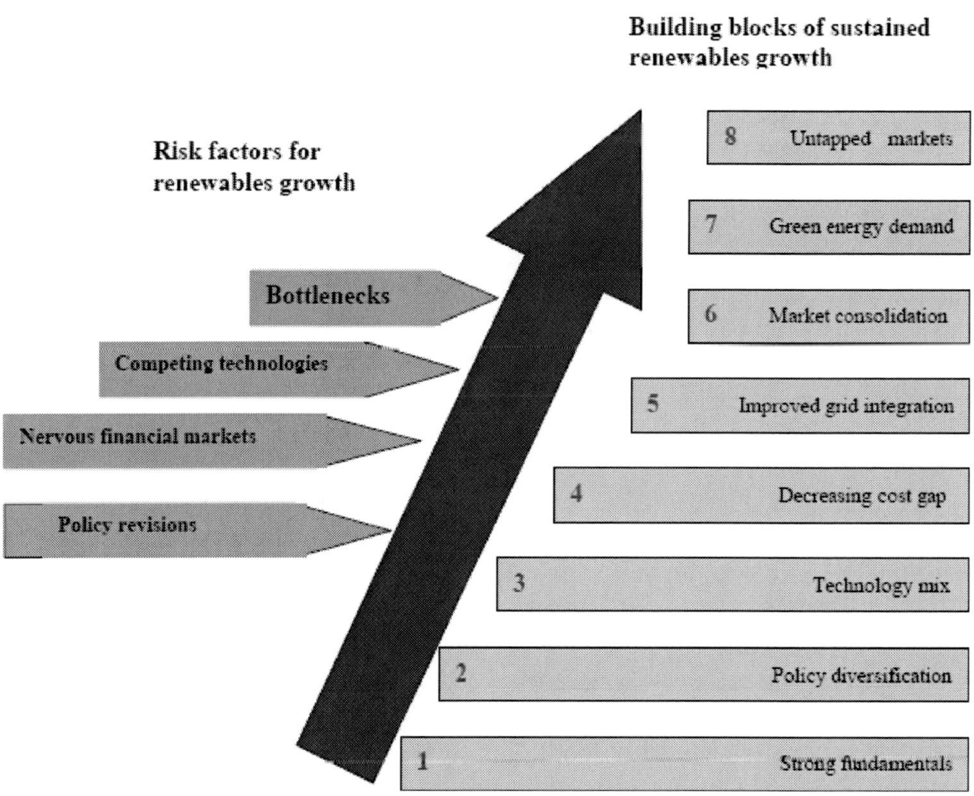

Figure 6. Summary of risk and risk mitigation factors.

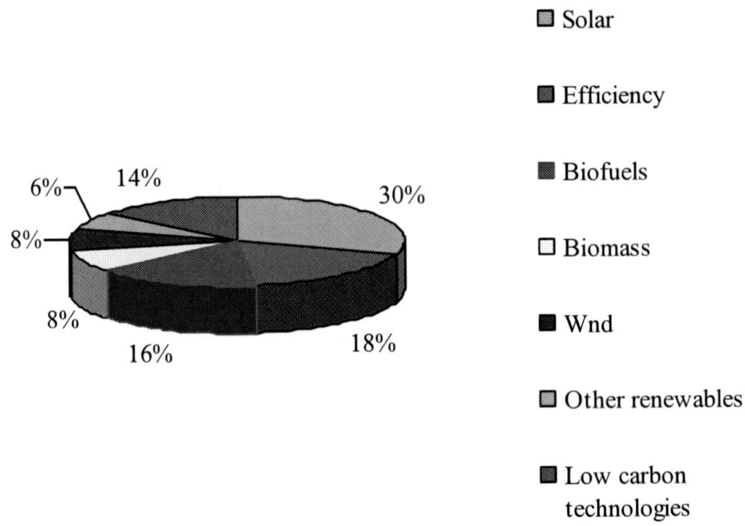

Figure 7. Investment in renewable and energy efficiency.

4. SOLAR ENERGY

Studies have been begun to estimate both the economic effects that climate change will have on global society as well as the costs of possible climate change mitigation and adaptation measures. Although the capacity to enact either a mitigation or adaptation strategy is based on country-specific conditions, technology and information availability, models have been used to calculate the approximate cost to stabilise atmospheric emissions at different levels.

Wind power is far from the only clean energy sector on the rise and many of the technologies following in its tracks are much more decentralised, including roof-top systems like photovoltaics (PVs) or solar thermal and energy efficiency technologies on the demand side. Solar and energy efficiency were actually the two largest sectors in terms of venture capital investment, with solar bringing in 30% and efficiency 18% (Figure 7). Besides the high level of early stage investment, mostly focusted on new technology development, these two sectors also fared well on the public stock markets, ranking second and third after wind. Solar would have overtaken wind on the public markets.

The potential of electric power generation from incorporating PVs in buildings is enormous. If the electrical power demand of many countries is to be supplemented by the use of PV, it is deemed necessary to integrate such systems into many of the building faces. Many larger structures such as superstores, public buildings and most houses use mass produced tiles on their roofs. Such areas lend themselves useful in contributing to the energy used in the building or to export to the electrical grid when active roof tiles (PV-tiles) are introduced as a part of the roof structure. The integration of PVs within both domestic and commercial roof offers the largest potential market for PV especially in the developed world. Numerous national programmes are attempting to stimulate this market using standard PV modules as a roof element. However, roofs are not static, uniform structures and so are not ideally suited to modules, which require precise, and planar mounting structures. To date, attempts at producing a PV roof tile, which accommodates current roof practice, in terms of both installation and aesthetics. The market for PV has historically been based on off-grid application where the relatively high cost of PV could be economically justified. In 2001 about 330 MWp of PV were produced and installed around the world and the growth rate of the industry is over 30% per year-100% per year in some countries with aggressive implementation schemes (Figure 8).

Several definitions of sustainable development have been put forth, including the following common one; development that meets the needs of the present without compromising the ability of future generations to meet their own needs. A recent World Energy Council (WEC) study found that without any change in our current practice, the world energy demand in 2020 would be 50-80% higher than 1990 levels. According to a recent USA Department of Energy (DoE) report, annual energy demand will increase from a current capacity of 363 million kilowatts to 750 million kilowatts by 2020. The world's energy consumption today is estimated to 22 billion kWh per year, 53 billion kWh by 2020. Such ever-increasing demand could place significant strain on the current energy infrastructure and potentially damage world environmental health by CO, CO_2, SO_2, NO_x effluent gas emissions and global warming. Achieving solutions to environmental problems that we face today requires long-term potential actions for sustainable development. In this regards, renewable energy resources appear to be the one of the most efficient and effective solutions since the intimate relationship between renewable energy and sustainable development. More rational use of energy is an important bridge to help transition from today's fossil fuel dominated world to a world powered by non-polluting fuels and advanced technologies such as photovoltaics (PVs) and fuel cells (FCs) [5].

Research into future alternatives has been and still being conducted aiming to solve the complex problems of this recent time, e.g., rising energy requirements of a rapidly and constantly growing world population and global environmental pollution. Therefore, options for a long-term and environmentally friendly energy supply have to be developed leading to the use of renewable sources (water, sun, wind, biomass, geothermal, hydrogen) and fuel cells. Renewables could shield a nation from the negative effect in the energy supply, price and related environment concerns. Hydrogen for fuel cells and the sun for PV have been considered for many years as a likely and eventual substitute for oil, gas, coal and uranium. They are the most abundant elements in the universe. The use of solar energy or PVs for the everyday electricity needs has distinct advantages to avoid consuming resources and degrading the environment through polluting emissions, oil spills and toxic by-products. A one-kilowatt PV system producing 150 kWh each month prevents 75 kg of fossil fuel from being mined.

There have now been a number of successful large-scale programmes of systems deployed for basic power needs in rural households in developing and less developed countries. Remote applications servicing other applications such as telecommunications, cathodic protection, water pumping, etc., continue to grow as well. The PV market will continue to grow strongly for the next several years at least, driven by incentive programmes, cost reductions and greater market awareness.

It is becoming evident, that the quality will be the key to the PV market (Figure 9). The European Commission's Altener Programme a Training Manual was developed by the Global Approval Programme for PVs (PV GAP) to help manufactures of PV products to introduce quality management in their production. The manual contain important up-dates of the PV

GAP Manuals for PV manufactures, published by the World Bank in 1999, including alignment with the 2000 edition of ISO 9001. This revised training manual was also translated from English into French, German and Spanish.

Water pumping is one of PV modalities that are growing in rural areas, mainly in developing countries. Due to the importance for health and food production it may be regarded as one of the noblest solar PV uses in isolated areas (Figures 10-11). Reliability and autonomy or self-reliance is understood as main factors for this high growth rates.

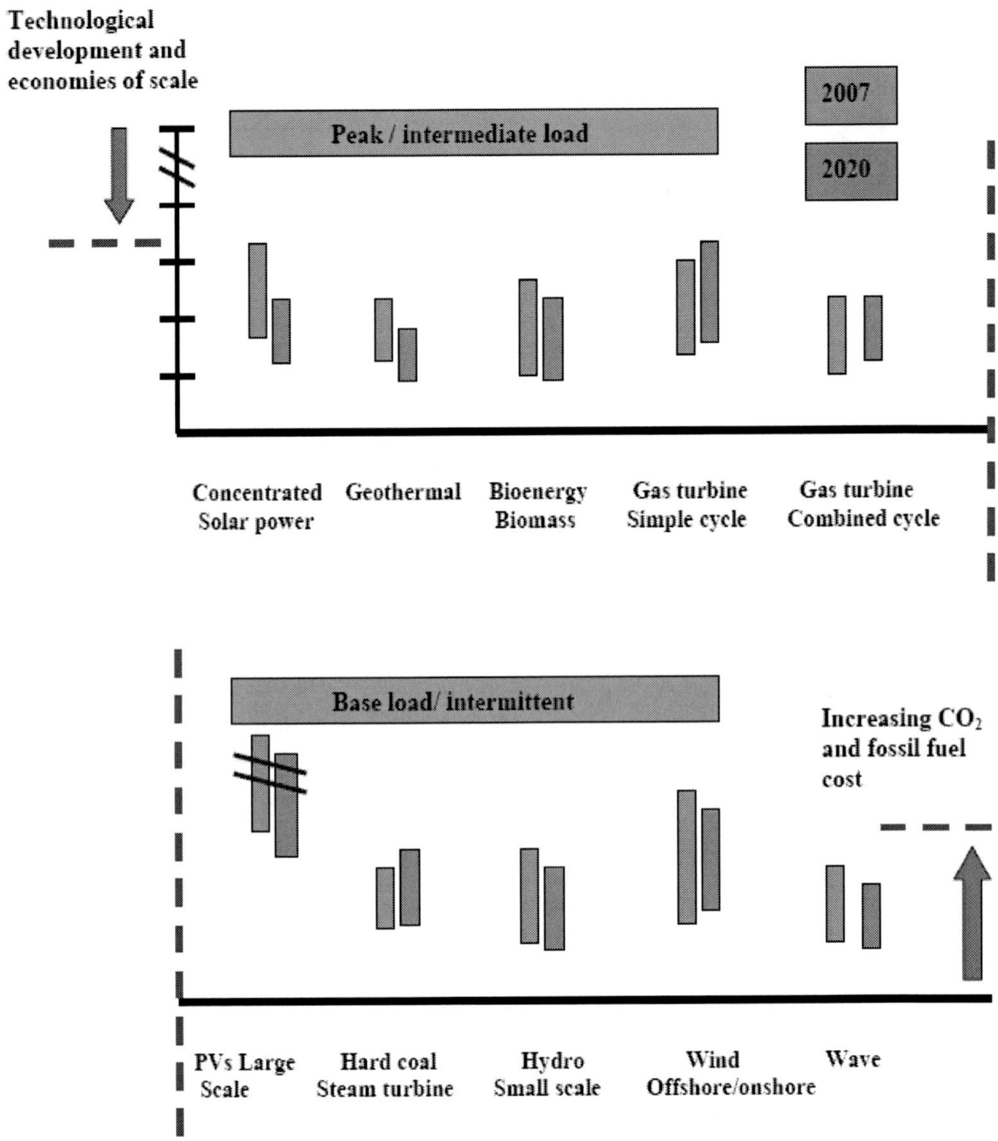

Figure 8. Cost increase for fossil primary energies and CO_2 emissions and continued technological development and economies of scale in renewables will improve competitiveness.

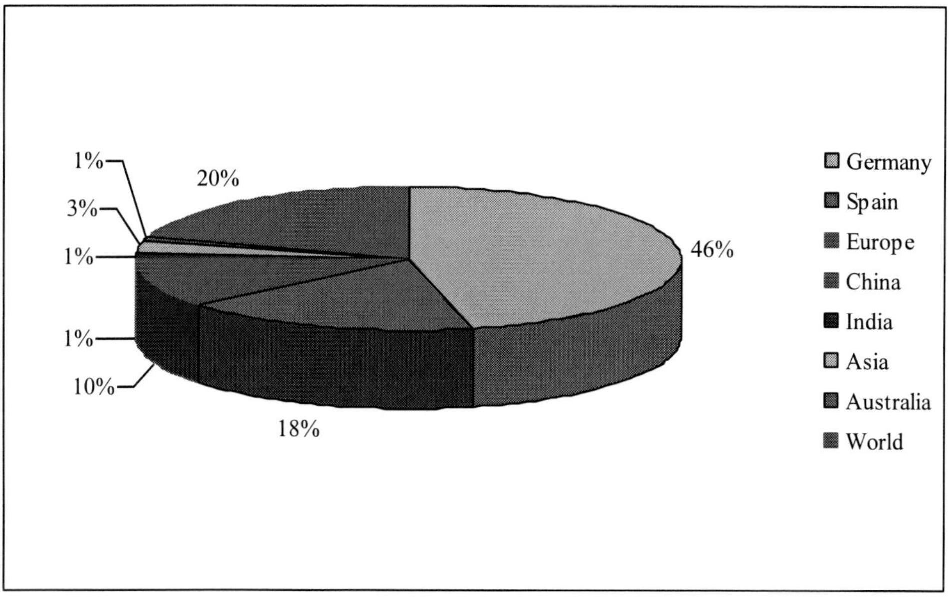

Figure 9. Differences in predicted PV market volumes worldwide until 2010.

5. Hydropower Potential

This section discusses various aspects of hydropower including harnessing ocean energy, hydroelectric dams and micro hydropower systems. This section on hydropower explores the factors associated with utilising the actual potential of hydropower energy. The section covers all the technological details, along with issues and challenges faced during the utilisation of hydropower energy. Major projects, power plants, players in the industry, the major role of the United States in the global hydropower industry, and the various environmental benefits of using hydropower energy are all explored in depth in this section.

The growing worldwide demand for renewable energy projects is being driven by ever increasing global energy consumption and the availability of carbon and renewable energy credits. Renewable energy is entering a new phase with additional funding becoming available from governments, from socially responsible equity funds, and from public capital raisings.

Hydropower is the capture of the energy derived from moving water for some useful purpose. Prior to the widespread availability of commercial electric power, hydropower was used for irrigation, milling of grain, textile manufacture, and the operation of sawmills.

Hydropower produces essentially no carbon dioxide or other harmful emissions. In contrast to burning fossil fuels, this energy is not a significant contributor to global warming through production of CO_2.

Hydroelectric power can be far less expensive than the electricity generated from fossil fuel or nuclear energy. Areas with abundant hydroelectric power attract industry. Environmental concerns about the effects of reservoirs may prohibit development of economic hydropower sources in some areas.

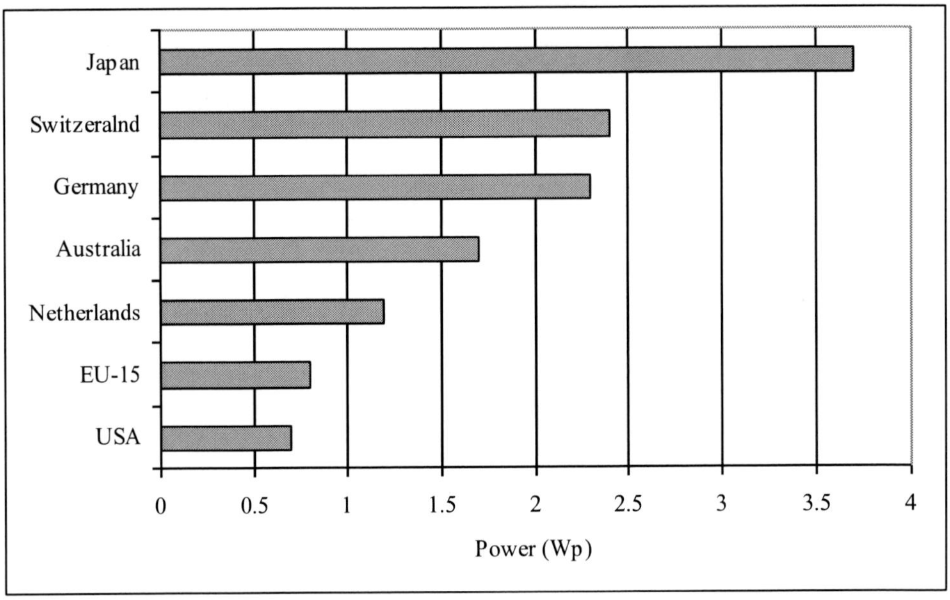

Figure 10. PV distributions for different countries per person.

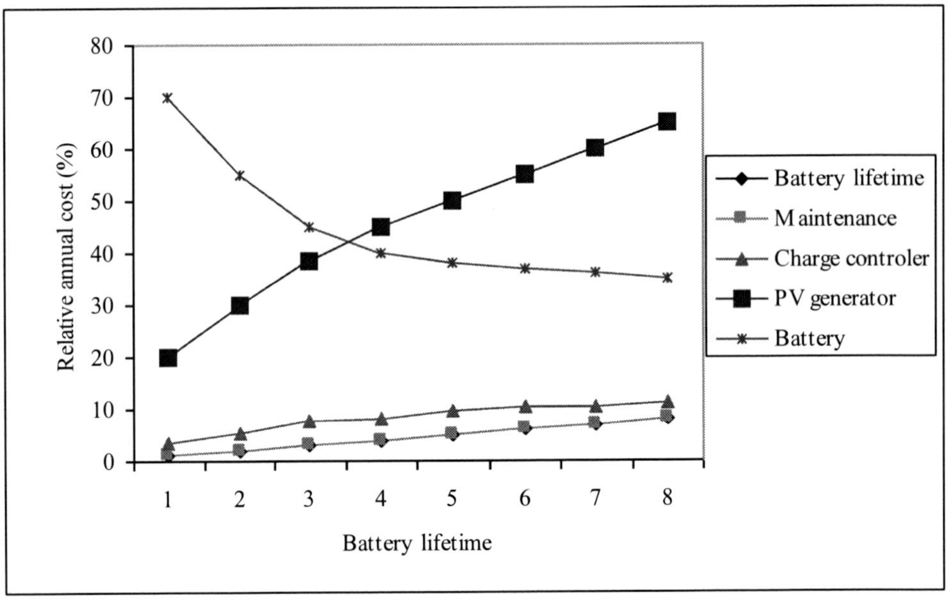

Figure 11. Relative annual costs of the components of small PV stand alone system with a lead-acid battery.

Hydropower currently accounts for approximately 20% of the world's electricity production, with about 650,000 MW installed and approximately 135,000 MW under construction or in the final planning stages. Notwithstanding this effort, there are large untapped resources on all continents, particularly in areas of the world that are likely to experience the greatest growth in power demand over the next century. It is estimated that only about a quarter of the economically exploitable water resources has been developed to date, leaving the potential for hydro to continue to play a large role in sustaining renewable global electricity production in the future.

Apart from a few countries with abundance, hydro power is normally applied to peak load demand because it can be readily stopped and started. Nevertheless, hydroelectric power is probably not a major option for the future of energy production in the developed nations, however, because most major sites within these nations are either already being exploited or are unavailable for other reasons, such as environmental considerations.

Future hydropower energy programmes must be put into practice in conjunction with sound policies that restrict the use of fossil fuels and natural resources and contribute to the reduction of emissions into the environment. Such a strategy should be based in a sound scientific basis, without ideology, politics or financial interests. It should be implemented on a worldwide basis and not limited to industrialised countries. To achieve this goal, existing hydropower energy options must be evaluated for implementation, new strategies must be formulated and new, innovative solutions have to be found.

All projects are required to have environmental impacts assessment conducted, covering all potential damage to the environment, mitigation and restoration, a reclamation plan including a resettlement programme for displaced residents, and the estimated implementation costs. All hydro projects are required to conduct an environmental impact study. Water is essential to industry for processes such as cooling, cleaning, diluting and sanitation. With increasingly stringent water abstraction limits, recent droughts and a growing interest in the environmental performance of businesses, there is a need for industry to reduce water use.

6. Gasification

An important consideration for operators of wastewater treatment plants (WWTPs) is how to handle the disposal of the residual sludge in a reliable, sustainable, legal and economical way. This by-product of wastewater treatment contains abundant organic material, including many kinds of bacteria. It also contains heavy metals and its composition is generally unknown. The benefits of drying sludge can be seen in two main treatment options:

- Use of the dewatered sludge as a fertiliser or in fertiliser blends.
- Incineration with energy recovery.

Use as a fertiliser option takes advantage of the high organic content 40-70% of the dewatered sludge and its high levels of phosphorous and other nutrients. However, there are a number of concerns about this route including:

- The chemical composition of the sludge (e.g., heavy metals, hormones and other pharmaceutical residues).
- Pathogen risk (e.g., salmonella, Escherichia coli, prionic proteins, etc.).
- Potential accumulation of heavy metals and other chemicals in the soil.

Table 5. Energy recovery from WWTP sludge

Form of sludge and usage	Cement kiln	Coal-fired power plant	Mono-incinerator	Refuse incinerator
Dewatered sludge	No	Yes	Yes/No*	Yes
Fuel substitute	-	No	No	No
Thermal dry biosolids	Yes	Yes	Yes	Yes
Substitute for fuel	Yes	Yes	Yes/No	No
Substitute for minerals	Yes	No	No	No
Type of process	Residue-free process	Fuel substitution	Disposal	Disposal

*Depending on type of plant.

Sludge can be applied as a fertiliser in three forms:

- Liquid sludge.
- Wet cake blended into compost.
- Dried granules.

Use as energy recovery option takes advantage of the energy available in the sludge's organic content. Drying the sludge reduces its water content, thus increasing its calorific value and making it easier to combust. It also reduces odours and improves handling, with lower transport and storage costs. Sludge from WWTPs is typically combusted in (Table 5):

- Cement kilns.
- Coal-fired power plants.
- Mono-incinerators, i.e., plants burning refuse-derived fuel or a single waste stream.
- Mixed waste incinerators, e.g., municipal waste incinerator.

The recognised advantages of energy recovery from sludge include:

- The high calorific value (similar to lignite) of dewatered sludge.
- The use of dewatered sludge as a carbon dioxide (CO_2) neutral substitute for primary fuels such as oil, gas and coal.
- The use of dewatered sludge is a 'sink' for pollutants such as heavy metals, toxic organic compounds and pharmaceutical residues, thus offering a potential disposal route for these substances provided the combustion plant has adequate flue gas cleaning.
- The potential, under certain circumstances, to utilise the inorganic residue from sludge incineration (incinerator ash), such as in cement or gravel.

The demands placed on the drying system are therefore critical and include:

- High process stability.
- High mechanical reliability.
- High safety standards under all operating conditions.

- Compliance with environmental legislation such as emission limits.
- A product with properties suitable for a wide range of uses.

The energy efficiency formula takes into account the energy generated by the plant and puts it in relation to the calorific value of the municipal waste [5]. The energy introduced into the process from outside (such as fossil fuels or electricity) is subtracted. The energy efficiency can be improved by, for instance, reducing the input of fossil fuels. According to the European commission's formula, the energy efficiency for waste-to-energy (WTE) plants is calculated as follows:

Energy efficiency = E_p- (E_f+E_i) / 0.97 (E_W+E_f) (1)

where:

E_p is the annual energy produced as heat or electricity in GJ/year. It is calculated with energy in the form of electricity multiplied by 2.6 and heat produced for commercial use multiplied by 1.1.

E_f is the annual energy input to the system in GJ/year from fuels contributing to the production of steam.

E_W is the annual energy in GJ/year contained in the treated waste calculated using the lowest net calorific value of the waste.

E_i is the annual energy imported in GJ/year, excluding E_W and E_f. For thermodynamic reasons, E_f must be deleted in the nominator of the equation, as it is included twice- in the nominator and the denominator.

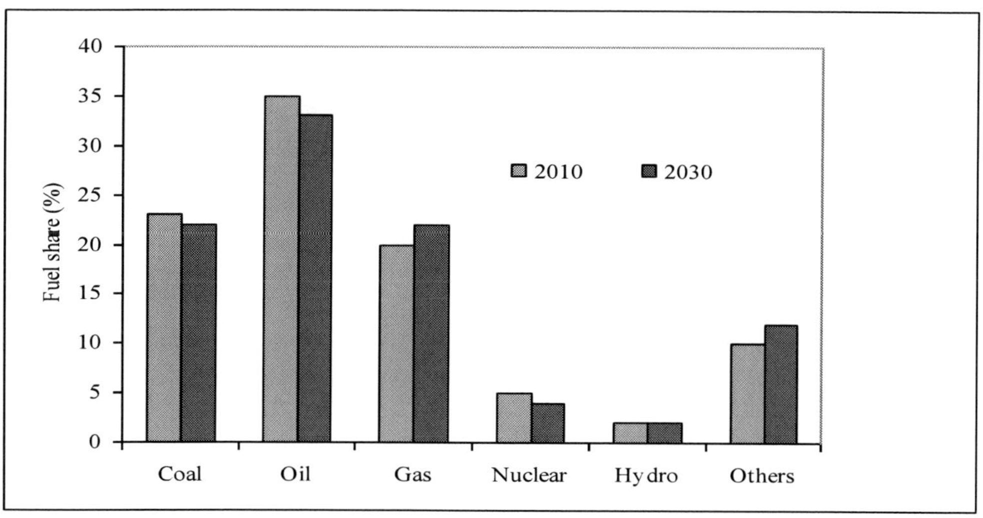

('Others' includes combustible renewables and waste, geothermal, wind and tidal energy).

Figure 12. Outlook for world total primary energy supply.

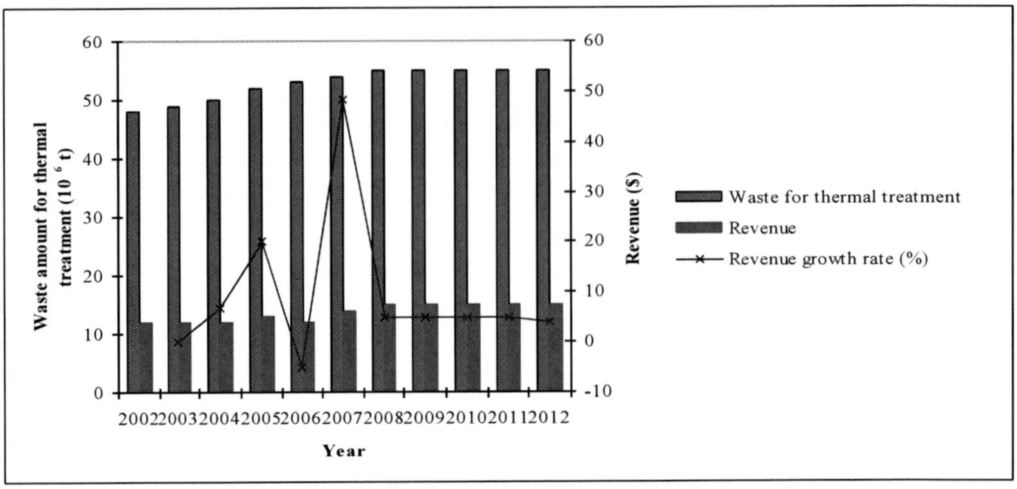

Figure 13. Volume shipments and revenue forecasts for the western European thermal waste treatment services.

An energy efficiency factor of 0.6, which has been proposed by commission, is too high for most existing plants. A threshold of 0.5 would be sufficient, with a further reduction of 0.1 for small plants and plants that produce electricity only due to a lack of demand for heat. A factor of 0.6 would disadvantage smaller plants as they generally need the same energy for operation as larger plants, but have a lower throughput. It should be noted, at this point, that the public tend to prefer smaller plants in order to reduce the distance waste is transported (Figures 12-13).

7. ENERGY RECOVERY

Mechanical-biological treatment (MBT) can be one option for improving the conservation of resources and energy in waste management systems. Mechanical-biological treatment enhances the conservation of embedded energy through recycling and allows potentially more efficient combustion or conversion of refuse-derived fuel (RDF). MBT encompasses a wide range of technologies aiming to process solid waste by a mixture of mechanical and biological separation. It also enables metals and other dry recyclables to be recovered (Figure 14). There are five main types of MBT process:

- Incorporating anaerobic digestion to generate biogas for electricity production. Anaerobic digestion also generates a digestate to be discharged or to be dewatered, producing a compost product.
- Producing an RDF product.
- Producing a compost product and/or a stabilised material for landfilling as well as a RDF product.
- Producing a compost product.
- Stabilising waste prior to landfill.

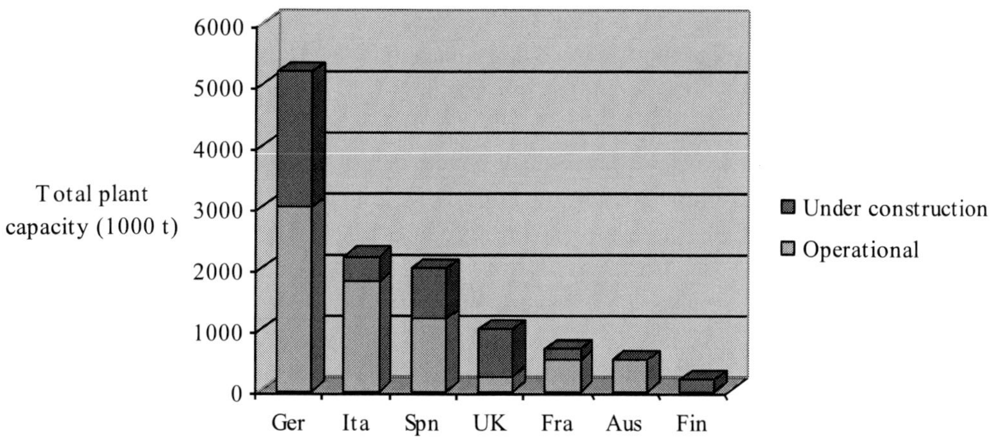

Figure 14. Capacity of plants in each country.

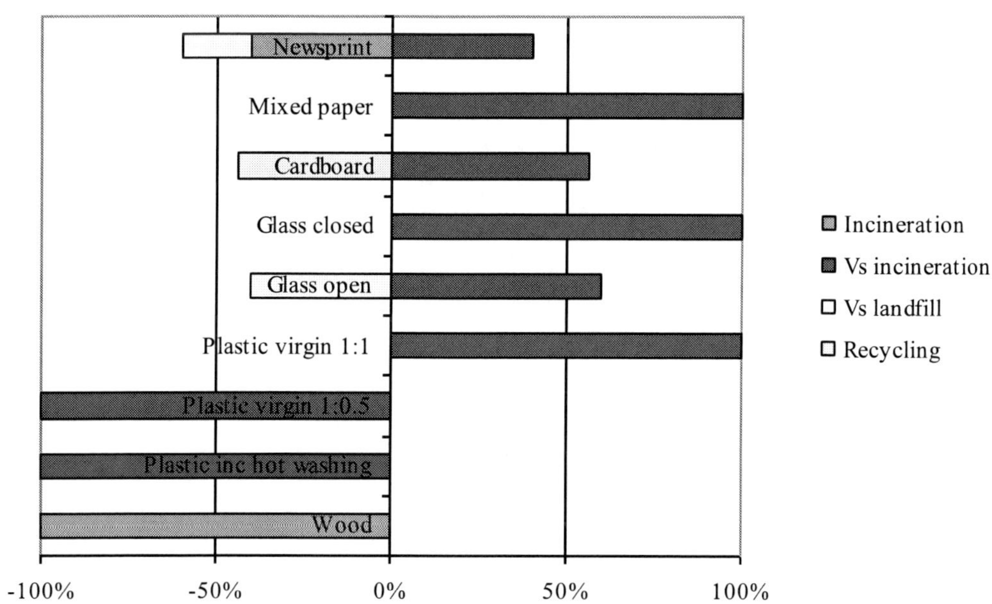

Figure 15. Performance of the life cycle analysis (LCA) on avoided greenhouse gas emissions.

Gas production in the gasification system is controlled in two ways:

- The system analyses the gas produced and proprietary software feeds instructions back to a control system, delivering constant management of operating variables.
- A surge and mixing tank blends the gas flow.

Social and environmental will benefit to the community from the utilisation of alternative energy and reduced fossil fuel consumption.

Recycling is hugely beneficial, both from an economic and environmental perspective. The environment and economy will both benefit from an improved and more efficient reverse supply chain. This, in turn, would result in increased and optimised recycling of waste streams that reduce the depletion of scarce natural resources- water, oil and minerals- while at the same time protecting the environment by keeping these valuable materials in service. Figure 15 shows how the individual LCAs performed on avoided greenhouse gas emissions in CO_2 equivalent by material [6]. A cornerstone of recycling is effective waste collection and many local authorities are now looking to households and businesses to ensure this is in place. It is no longer sufficient to leave one or more bags of 'rubbish' at the end of a drive or back of a workplace.

Odorous emissions are a serious concern relating to biowaste treatment facilities and affect the likelihood of planning permission being granted particularly in urban areas. Research has shown odour is one of the prime concerns from urban residents where MBT plants are proposed. An often-unrecognised aspect of the waste management field is the many examples of individuals giving back. A conventional biofilter operates by engineering the correct environment in a closed vessel, ensuring that the bacteria within it are cultivated to effectively biodegrade the compound in the air that passes through it. Recirculation of the airstreams improves the rate of degradation by the bacteria. This has several advantageous characteristics:

- A surface s structure that is excellent for supporting bacteria and has the ability to retain moisture in the event of temporary water failure.
- A self-supporting structure with good packing characteristics that ensures there is minimal pressure drop across the media bed (thus reducing running costs).
- The right conditions to encourage bacteria to metabolise more than grow, which minimises sludge production and hence associated disposal.

8. Geothermal Heat

This study explores the factors associated with utilising the actual potential of geothermal energy. It also covers all the technological details, along with issues and challenges faced during the utilisation of geothermal energy. Major projects, power plants, players in the industry, the major role of the United States in the global geothermal industry, the active role of the US Department of Energy, and the various environmental benefits of using geothermal energy are all explored in-depth in this study.

Geothermal power is the use of geothermal heat to generate electricity. Geothermal comes from the Greek words geo, meaning earth, and therme, meaning heat.

The utilisation of geothermal energy for the production of electricity dates back to the early part of the twentieth century. For 50 years the generation of electricity from geothermal energy was confined to Italy and interest in this technology was slow to spread elsewhere. In 1943 the use of geothermal hot water was pioneered in Iceland. Estimates of exploitable worldwide geothermal energy resources vary considerably.

The largest dry steam field in the world is the Geysers, about 90 miles (145 km) north of San Francisco. The Geysers began in 1960, which has 1360 MW of installed capacity and produces about 1000 MW net. Calpine Corporation now owns 19 of the 21 plants in the Geysers and is currently the United States' largest producer of renewable geothermal energy.

The Northern California Power Agency and Santa Clara Electric own the other two plants jointly. Since the activities of one geothermal plant affects those nearby, the consolidation plant ownership at the Geysers has been beneficial because the plants operate cooperatively instead of in their own short-term interest. Injecting treated sewage effluent from the City of Santa Rosa and the Lake County sewage treatment plant now recharges the Geysers. This sewage effluent used to be dumped into rivers and streams and is now piped to the geothermal field where it replenishes the steam produced for power generation.

Another major geothermal area is located in south central California, on the southeast side of the Salton Sea, near the cities of Niland and Calipatria, California. As of 2001, there were 15 geothermal plants producing electricity in the area. CalEnergy owns about half of them and various companies own the rest. Combined the plants have a capacity of about 570 megawatts. Geothermal energy can be used as an efficient heat source in small end-use applications such as greenhouses, but the consumers have to be located close to the source of heat. Geothermal energy has a major environmental benefit because it offsets air pollution that would have been produced if fossil fuels were the energy source. Geothermal energy has a very minor impact on the soil - the few acres used look like a small light-industry building complex. Since the slightly cooler water is reinjected into the ground, there is only a minor impact, except if there is a natural geyser field closed by. The world's decentralised energy industry is right to look towards developing countries as potential markets for its wares. These countries have the opportunity to miss out the industrial development stage characterised by a single, centralised electricity transmission and distribution system, in favour of a hybrid of local grids liberally supplied by small-scale decentralised power generators.

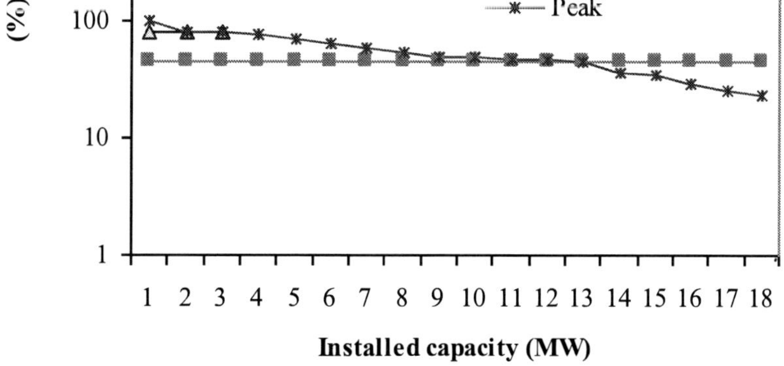

Figure 16. Load duration curve showing the base, intermediate, peaking load segments, and necessary contingency reserve.

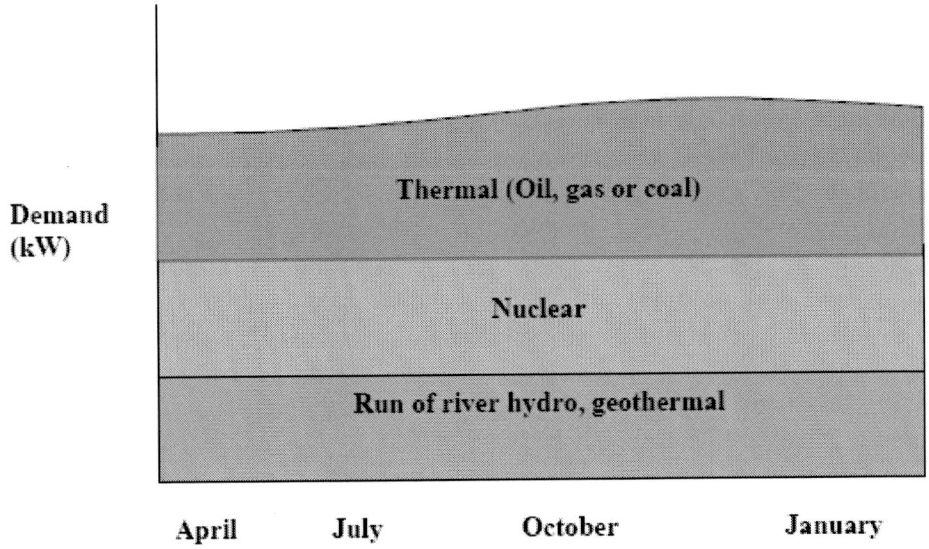

Figure 17. Image of annual load profile of power plants.

9. Greenhouse Gas Emissions

The typical challenge in the developing world is lack of everything, including electricity. When striving to feed the increasing power demand to enable social reform and industrial growth, local decision makers face the question of which route to take. Shall we copy the model of the rich countries, or could there be another, maybe better way? The need to reduce CO_2 emissions presents a new, additional challenge, difficult even for the richest of nations.

Conventional, centralised electricity networks are the norm in the developed world. However, the present energy infrastructure of the developed countries was mainly created during the monopolistic utility era of the past. The utilities had the power to decide what kind of capacity to construct and how to construct the grid. There was practically no competition. The main challenge for many utilities was to get construction permits for building new generation capacity- the construction itself was practically risk free as they could turn their cost structure into a solid power tariff. Capital was easily available and cost competitiveness and overall systems cost optimisation were not the main concerns.

Today the market situation and rules have permanently changed. Progressive modern utilities have left the past behind and are striving towards a modern, competitive energy system. However, one remainder of the past, still to some extent maintaining the economies of scale thinking, is to look at and calculate power plant and grid investments separately, as if they had nothing, or very little, to do with each other. Installing flexible, high-efficiency peaking and grid stability generation capacity closer to the load pockets of cities and industrial areas would be a natural way to go, but it does not fit the model of old, large-scale utility thinking (Figure 16). A typical electricity system built by a rich country monopoly utility 'wasted' capital in the following ways:

- Large, steam-fired power plants, running on part load, were used for frequency and load control (unless hydro non-dispatchable wind condition, but the problem has been solved with the very strong grid, which can transmit the wind power over long distances to remote countries.

- Efficient peaking capacity was not constructed; instead, the so-called peaking plants were typically based on largest possible simple cycle industrial gas turbines located in critical points in grid. This capacity functional mainly as an emergency reserve at grid nodes and was hardly ever used as it has such poor heat rate and relatively long starting time.
- Excessive base-load capacity was constructed. This was possible as there was an ensured return on asset investments.
- The grid has been sized to transmit the full peak power from large remote power plants to the consumption centres in cities.

Utilities in the developing world face a major challenge in developing their electricity systems. The parameters to optimise at the same time are:

- Reliability of supply.
- System flexibility and preparedness for load growth.
- Economical competitiveness, i.e., cost effectiveness, and elimination of 'wastes'.
- Access to power for the whole nation.
- Ensured access to fuels and fuel flexibility.

Combined heat and power (CHP) has been installed and used for many years as a highly efficient energy supply system to fulfil electricity and thermal energy needs in a range of applications around the world. As environmental issues, particularly those of climate change caused by greenhouse gases (GHGs) become as an effective means to reduce emissions is more pronounced than ever. However, despite CHP's significant contribution to the reduction of CO_2 emissions, calculation of its effects can result in preventing further dissemination of CHP. This is because, when an appropriate calculation method is used, there is a risk of under-estimating the effects of CHP and even, in some instances, estimating erroneously an increase in CO_2 emissions as a result of installing a CHP system. Proper estimation of the benefits of CHP will be vital in encouraging more businesses to install CHP systems.

Considerable debate surrounds the calculation of CO_2 emissions as a result of the use of the two emission factors [6]:

- All fuels average emission factor (AEF) - used to calculate emissions from power consumption.
- Marginal emission factor (MEF) - used to estimate reduced amount of CO_2 emissions as a result of lower power consumption.

CO_2 emission by consumption of a fuel is calculated as follows:

Emission amount (kg CO_2) = emission factor of a fuel (kg CO_2/MJ, kWh, etc.) x energy consumed (MJ, kWh, etc.) (2)

In the case of electricity, however, CO_2 is not emitted at the point of use. A mix of fuels – some CO2 – emitting plants such as oil, coal and natural gas-based thermal power plants and non-CO2 – emitting power sources such as nuclear, hydro and renewables, usually generates the power on the grid. As a result, when calculating CO_2 emission by the use of grid power, the use of AEF or MEF is preferred. Therefore, CO_2 emission by an electricity user is estimated as follows:

CO_2 emission volume (kg CO_2) = AEF (kg CO_2/kWh) x power consumed (kWh) (3)

CO_2 emission of grid electricity is not accounted for on the demand side, rather on the generation side. AEF is obtained by dividing the total CO_2 emission volume from thermal power plants connected to the grid by the total power supplied through the grid including those that are free of emissions, indicating average CO_2 emission per kWh of power on grid.

AEF (kg CO_2/kWh) = total CO_2 emissions from all thermal plants on grid (kg/ CO_2)/total power supplied to grid from all sources (kWh) (4)

To calculate the amount of CO_2 emissions reduced because of lower power consumption, the emission factor for these plants must be multiplied by the reduced power consumption. The factor used is called the marginal emission factor or MEF. The reduced emission amount is obtained by the following calculation:

Emission reduction (kg CO_2) = MEF (kg CO_2/kWh) x electricity conserved (kWh) (5)

10. Energy Efficiency

Eventually renewable energies will dominate the world's energy supply system. There is no real alternative. Humankind cannot indefinitely continue to base its life on the consumption of finite energy resources. Today, the world's energy supply is largely based on fossil fuels and nuclear power. These sources of energy will not last forever and have proven to be contributors to our environmental problems. The environmental impacts of energy use are not new but they are increasingly well known; they range from deforestation to local and global pollution. In less than three centuries since the industrial revolution, humankind has already burned roughly half of the fossil fuels that accumulated under the earth's surface over hundreds of millions of years. Nuclear power is also based on a limited resource (uranium) and the use of nuclear power creates such incalculable risks that nuclear power plants cannot be insured.

Renewable sources of energy are an essential part of an overall strategy of sustainable development. They help reduce dependence of energy imports, thereby ensuring a sustainable supply. Furthermore renewable energy sources can help improve the competitiveness of industries over the long run and have a positive impact on regional development and employment. Renewable energy technologies are suitable for off-grid services, serving those in remote areas of the world without requiring expensive and complicated grid infrastructure. In 2007, the United States outlined plans to ease out of its foreign oil dependence through the use of renewable energy resources, and reduce gas usage by a full 20% in ten years through alternative fuels. Extending hope and opportunity depends on a stable supply of energy that keeps nation's economy running and environment clean.

Many wind/solar farms are located in remote areas served by low capacity radial distribution networks. These networks often have insufficient capacity to ship the power to demand centres when the turbines are generating at full capacity.

The public awareness of the depletion of the ozone layer has increased since 1970s. The ozone layer in the stratosphere protects life on earth against the ultraviolet radiation from the sun. Scientists and politicians argued for some years on the reasons for the reduced ozone layer, which was recognised especially over the Antarctic. The fact is that the ozone is depleted by the presence of chlorine. Further, very low temperatures as well as sunlight are required for having the process running. Vapour compression refrigeration systems using fluorocarbons, hydrocarbons or ammonia represent the established technology for household, commercial and industrial refrigeration and air-conditioning.

Understanding the earth and the processes that shape it is fundamental to the successful development and sustainable management of our planet. The evolution of the earth's crust over geological time has resulted in diverse, often beautiful, landscapes formed by earthquakes, oceans, fire and ice. These landscapes are also a source of mineral wealth and water, a base for engineering projects and a receptacle for the waste. Minerals are vital for manufacturing, construction, energy generation and agriculture. Some of the requirements can be met by increasing the use of recycled and renewable resources, but the need for new mineral resources continues. Minerals are important to maintaining the modern economy and lifestyle. Hence, everyone must make best use of these valuable assets whilst minimising the impact of their extraction on the environment. Global energy use will rise dramatically but world oil and gas production will eventually decline. Geological hazards account for huge loss of life and damage to property. Poorer countries often suffer the most due to inappropriate land-use planning and in the future climate change may worsen these effects. Clean drinking water is a basic human need. The environmental sustainability of water resources, especially when balancing ecological and human needs with economic growth, is of major concern to governments worldwide. Increases in the world's population, the growth of cities, industrial development and ever-increasing waste and pollution can increase pressure on the environment. Understanding the effects of climate change is a key issuc for society and researches focus on how to predict future climate change events and how to mitigate them.

Two essential components for economic development are a soundly based knowledge of a country's natural resources, such as groundwater, minerals and energy, and an understanding of the geological environment. The latter includes the potential effects of earthquakes, volcanic eruptions, landslides and the forces of erosion and deposition on rivers and coastlines. Helping developing countries acquire this knowledge and apply it to promote economic growth, sustainable livelihoods and the protection of people. Sustainable land use involves protection of the natural environment (viable agriculture, forestry, water resources, and soil functions), the quality and character of the countryside and existing communities. Groundwater is an essential source of drinking water (70%). Although commonly regarded as pure, it is often vulnerable to pollution or may contain natural concentrations of elements that can adversely affect health. Population growth, intensive agriculture, urbanisation and higher standards of living have resulted in increasing, often unsustainable, demands on groundwater resources. These pressures and the likely effects of climate change mean that groundwater systems must be better understood so that they can be evaluated effectively, managed sustainably and protected securely.

District heating plays a very important role in transition economies. District heating systems in transition economies often face financial, technical or managerial problems largely created by an inadequate policy framework. These challenges include lack of customer focus, low efficiency, excess capacity, corruption and an uneven playing field. Figure 18 shows how these challenges are interrelated and create a vicious circle that undermines finances and competitiveness of district heating companies, and jeopardising their long-term sustainability.

10.1. Energy Supply

The earth is believed to be close to a state of thermal equilibrium where the energy, which is received at the surface by solar radiation, is lost again at night and the much smaller amount of energy, which is generated by the decay of unstable isotopes of Uranium, Thorium and Potassium distributed within the earth is balanced by the small continuous heat flux from the earth's interior to the pecans and atmosphere.

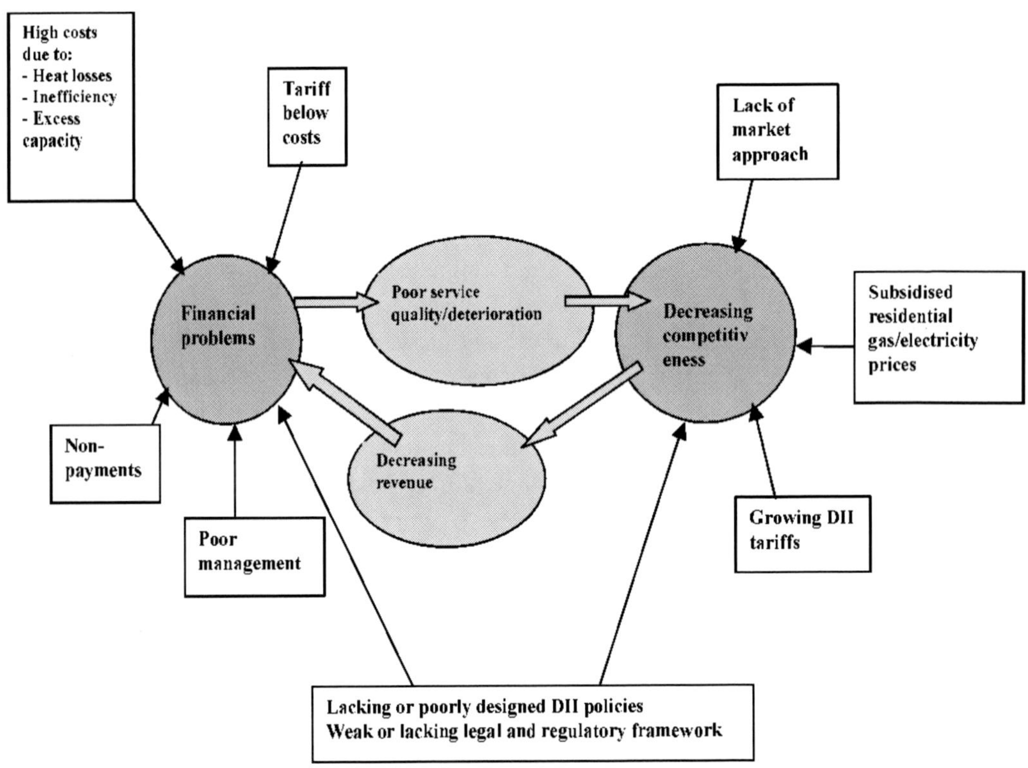

Figure 18. Key challenges for district heating systems in transition economies.

The use of geothermal energy involves the extraction of heat from rocks in the outer part of the earth. It is relatively unusual for the rocks to be sufficiently hot at shallow depth for this to be economically attractive. Virtually all the areas of present geothermal interest are concentrated along the margins of the major tectonic plates, which form the surface of the earth. Heat is conventionally extracted by the forced or natural circulation of water through permeable hot rock.

There are various practical difficulties and disadvantages associated with the use of geothermal power:

- Transmission: geothermal power has to be used where it is found. In Iceland it has proved feasible to pipe hot water 20 km in insulated pipes but much shorter distances are preferred.
- Environmental problems: these are somewhat variable and are usually not great. Perhaps the most serious is the disposal of warm high salinity water where it cannot be reinjected or purified. Dry steam plants tend to be very noisy and there is releases of small amounts of methane, hydrogen, nitrogen, amonia and hydrogen sulphide and of these the latter presents the main problem.

The geothermal fluid is often highly chemically corrosive or physically abrassive as the result of the entrained solid matter it carries. This may entail special plant design problems and unusually short operational lives for both the holes and the installations they serve.

Because the useful rate of heat extraction from a geothermal field is in nearly all cases much higher than the rate of conduction into the field from the underlying rocks, the mean temperatures of the field is likely to fall during exploitation. In some low rainfall areas there may also be a problem of fluid depletion. Ideally, as much as possible of the geothermal fluid should be reinjected into the field. However, this may involve the heavy capital costs of large condensation installations. Occasionally, the salinity of the fluid available for reinjection may be so high (as a result of concentration by boiling) that is unsuitable for reinjection into ground. Ocasionally, the impurities can be precipitated and used but this has not generally proved commercially attractive.

10.2. Refrigeration

The refrigeration industry has had to face environmental challenges. Regulation and behaviour related to the ozone layer differ from one country to another. Problems related to global warming are likely to cause similar problems in very different ways. These two kinds of problems may lead company managers to change/modernise equipment. These environmental factors must not hide other important development factors such as those related to hygiene, organoleptic quality, control methods and equipment packaging, etc. All equipment change and investment projects must take all of them into consideration.

The more developed a country, is the more widely refrigeration is used. Food preservation is still the main use of refrigeration, followed by air conditioning, energy savings and transport (heat pump and liquefied gases), industrial processes and medicine. When consumers demand more and more fresh products, handling such products becomes increasingly difficult: cold stores, display cabinets, refrigerated trucks have to be much more effective. Difficulties are mostly related to:

- Very narrow ranges, in many cases, between temperatures involving microbial or chemical risk and temperatures that cause chilling injury or freezing.
- No or little overlapping between temperature ranges permitted for different products.
- No or little possibility for the consumer or the seller to evaluate the remaining life spans of the product or even to perceive any risk.
- Humidity control, vapour pressure differences have serious consequences, not only on the weight but also on the unit value of products.

The production and use of Chloro Fluoro Carbon (CFC) refrigerants have been phased out in developed countries. Hydro Chloro Fluoro Carbon (HCFC) refrigerants are now subject to regulation and are scheduled to be phased out in the twenty-first century. Recent developments in refrigeration technology and environment (climate) control have provided a better quality of life for mankind. As is known, refrigeration technology has played a key role in preserving the quality of food and eliminating waste (spoilage). The refrigerated transport of perishable products provides a critical link in the cold chain between the consumer and the producers, processors, shippers, distributors and retailers. The transport refrigeration industry incorporated mechanical vapour compression mechanisms into the first transport refrigeration units. Therefore, the physical, chemical and thermodynamic properties of these refrigerants had a major impact on the design of compressors and other refrigeration system components. Many refrigerants have excellent thermodynamic properties suitable for use in vapour compression refrigeration systems, but are limited by other properties such as toxicity, flammability and chemical stability within the refrigeration system.

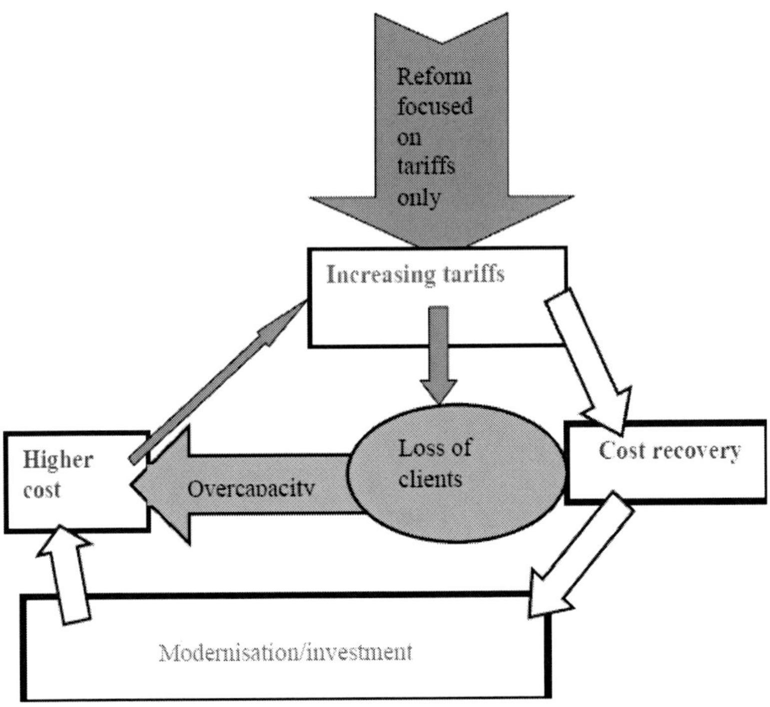

Figure 19. Unsustainable tariff growth.

Most district heating systems in former planned economies are less efficient and oversized. In other words, their supply infrastructure is larger than necessary to meet current demand. This problem can be exacerbated when they lose customers. Reforms, particularly for tariffs, are needed to improve finances and break the vicious circle of deteriorating competitiveness. Yet policymakers need to design the reforms very carefully for them not to backfire and make matters worse, as illustrates in Figure 19.

Figure 20. A photograph of a laying ground-loop heat exchanger.

Figure 21. A photograph shows the earth loop vapour and liquid.

Table 6. Values of geothermal parameters

Parameter	Lower	Average	Upper
q (HFU)	0.8	1.5	3.0 (non volcanic) ≈100 (volcanic)
k =cal cm^{-2} sec^{-1} $^{o}C^{-1}$	$2x10^{-3}$	$6x10^{-3}$	$12x10^{-3}$
β =°C/km	8	20	60 (non volcanic) ≈300 (volcanic)

10.3. Temperature Distributions

World capacity of geothermal energy is growing at a rate of 2.5% per year from a 2005 level of 28.3 GW [6]. The GSHPs account for approximately 54% of this capacity almost all of it in North America and Europe [6]. The involvement of the UK is minimal with less than 0.04% of world capacity and yet is committed to substantial reduction in carbon emission beyond the 12.5% Kyoto obligation to be achieved by 2012. The GSHPs offer a significant potential for carbon reduction and it is therefore expected that the market for these systems will rise sharply in the UK in the immediate years ahead given to low capacity base at present. There are numerous ways of harnessing low-grade heat from the ground for use as a heat pump source or air conditioning sink. For small applications (residences and small commercial buildings) horizontal ground loop heat exchangers buried typically at between 1 m and 1.8 m below the surface can be used provided that a significant availability of land surrounding the building can be exploited which tends to limit these applications to rural settings. Horizontal ground loop heat exchangers can be used to circulate refrigerant (direct heat exchanger) or a water/antifreeze mixture (indirect heat exchange) and rely to some extent on solar input during summer for earth temperature recovery following winter heat extraction (Figures 20-21). For a more economical use of available land, a vertical ground loop heat exchange array or ''borehole'' array can be used typically involving a matrix of vertical borehole heat exchangers spaced at 5 m at depths of up to 180 m. High density plastic (typically high density polyethylene) tube of 20-40 mm nominal diameter is fed down to 100-150 mm diameter borehole to form a U-tube with the borehole subsequently grouted using a high conductivity hard-setting compound (usually bentonite). These and others methods of low grade-ground heat harvesting have been reviewed in details from the perspective of UK application by Rawlings [7].

Heat generation within the earth is approximately 2700 GW, roughly an order of magnitude greater than the energy associated with the tides but about four orders less than that received by the earth from the sun [8].

Temperature distributions within the earth depend on:

- The abundance and distribution of heat producing elements within the earth.
- The mean surface temperature (which is controlled by the ocean/atmosphere system).
- The thermal properties of the earth's interior and their lateral and radial variation.
- Any movements of fluid or solid rock materials occurring at rates of more than a few millimetres per year.

Of these four factors, the first two are of less importance from the point of view of geothermal energy. Mean surface temperatures range between 0-30°C and this variation has a small effect on the useable enthalpy of any flows of hot water. Although radiogenic heat production in rocks may vary by three orders of magnitude, there is much less variation from place to place in the integrated heat production with depth. The latter factors, however, are of great importance and show a wide range of variation. Their importance is clear from the relationship:

$$\beta = q/k \tag{6}$$

where:

β is the thermal gradient for a steady state (°C/km), q is the heat flux (10^{-6} cal cm^{-2} sec^{-1}) and k is the thermal conductivity (cal cm^{-1} sec^{-1} $°C^{-1}$).

The first requirement of any potential geothermal source region is that β being large, i.e., that high rock temperatures occur at shallow depth. Beta will be large if either q is large or k is small or both. By comparison with most everyday materials, rocks are poor conductors of heat and values of conductivity may vary from 2 x 10^{-3} to 10^{-2} cal cm^{-1} sec^{-1} $°C^{-1}$. The mean surface heat flux from the earth is about 1.5 heat flow units (1 HFU = 10^{-6} cal cm^{-2} sec^{-1}) [8]. Rocks are also very slow respond to any temperature change to which they are exposed, i.e., they have a low thermal diffusivity:

$$K = k/\rho C_p \tag{7}$$

where:

K is thermal diffusivity; ρ and C_p are density and specific heat respectively.

These values are simple intended to give a general idea of the normal range of geothermal parameters (Table 6). In volcanic regions, in particular, both q and β can vary considerably and the upper values given are somewhat nominal.

In many geothermal areas natural heat exchangers exist in the form of large-scale, sub-surface water circulation systems, which commonly give rise to hot springs or Geyser activity at the surface (Table 7).

10.4. Thermodynamic Analysis of Refrigeration Cycles

Thermodynamics is the study of energy, its transformations and its relation to states of matter. A thermodynamic system is a region in space or a quantity of matter bounded by a closed surface. The surroundings include everything external to the system and the system is separated from the surroundings by the system boundaries. These boundaries can be movable or fixed, real or imaginary.

Potential energy (PE) is caused by attractive forces existing between molecules, or the elevation of system.

$$PE = mgz \tag{8}$$

where

m is the mass

g is local acceleration of gravity

z is the elevation above horizontal reference plane

Net amount of energy added to system] = [Net increase of stored energy in system (9)

or

[Energy in] – [Energy out] = [Increase of stored energy in system] (10)

Refrigeration cycles transfer thermal energy from a region of low temperature T_R to one of higher temperature. Usually the higher temperature heat sink is the ambient air or cooling water, at temperature T_o, the temperature of the surroundings.

Performance of a refrigeration cycle is usually described by a coefficient of performance (COP), defined as the benefit of the cycle (amount of heat removed) divided by the required energy input to operate the cycle:

COP = useful refrigerating effect / Net energy supplied from external sources (11)

For a mechanical vapour compression system, the net energy supplied is usually in the form of work, mechanical or electrical and may include work to the compressor and fans or pumps. Thus,

COP = Qevap / Wnet (12)

Table 7. Selected comparative cost data for geothermal energy [8]

Geothermal field	Geothermal production	Local average, other fuel
Electricity, US mills/kWh		
Iceland	2.5-3.5	-
Italy	4.8-6.0	~7.5
Japan	4.6	~6.0
Mexico	4.1-4.9	~8.0
Russia	7.2	~10.0
USA	5.0	7.0
Space heating, US$/Gcal energy		
Iceland	4.0	6.7
Hungary	3.0	11.0
Refrigeration, US$/Gcal		
New Zealand	0.12	2.40
Drying diatomite, US$/ton		
Iceland	~2	~12

10.5. Built Environment

The heating or cooling of a space to maintain thermal comfort is a highly energy intensive process accounting for as much as 60-70% of total energy use in non-industrial buildings. Of this, approximately 30-50% is lost through ventilation and air infiltration. However, estimation of energy impact of ventilation relies on detailed knowledge about air change rate and the difference in enthalpy between the incoming and outgoing air streams. In practice, this is a difficult exercise to undertake since there is much uncertainty about the value of these parameters [9]. As a result, a suitable datum from which strategic planning for improving the energy efficiency of ventilation can be developed has proved difficult to establish [10]. Efforts to overcome these difficulties are progressing in the following two ways:

- Identifying ventilation rates in a representative cross section of buildings.
- The energy impact of air change in both commercial and domestic buildings.

In addition to conditioning energy, the fan energy needed to provide mechanical ventilation can make a significant further contribution to energy demand. Much depends on the efficiency of design, both in relation to the performance of fans themselves and to the resistance to flow arising from the associated ductwork. Figure 22 illustrates the typical fan and thermal conditioning needs for a variety of ventilation rates and climate conditions.

The building sector is an important part of the energy picture. Note that the major function of buildings is to provide an acceptable indoor environment, which allows occupants to carry out various activities. Hence, the purpose behind this energy consumption is to provide a variety of building services, which include weather protection, storage, communications, thermal comfort, facilities of daily living, aesthetics, work environment, etc. However, the three main energy-related building services are space conditioning (for thermal comfort), lighting (for visual comfort), and ventilation (for indoor air quality). Pollution-free environments are a practical impossibility. Therefore, it is often useful to differentiate between unavoidable pollutants over which little source control is possible, and avoidable pollutants for which control is possible. Unavoidable pollutants are primarily those emitted by metabolism and those arising from the essential activities of occupants. 'Whole building' ventilation usually provides an effective measure to deal with the unavoidable emissions, whereas 'source control' is the preferred and sometimes only practical, method to address avoidable pollutant sources [11]. Hence, achieving optimum indoor air quality relies on an integrated approach to the removal and control of pollutants using engineering judgment based on source control, filtration, and ventilation. Regardless of the kind of building involved, good indoor air quality requires attention to both source control and ventilation. While there are sources common to many kinds of buildings, buildings focusing on renewable energy may have some unique sources and, therefore, may require special attention [11]. In smaller (i.e., house size) buildings, renewable sources are already the primary mechanism for providing ventilation. Infiltration and natural ventilation are the predominant mechanisms for providing residential ventilation for these smaller buildings.

Ventilation is the building service most associated with controlling the indoor air quality to provide a healthy and comfortable environment. In large buildings ventilation is normally supplied through mechanical systems, but in smaller ones, such as single-family homes, it is principally supplied by leakage through the building envelope, i.e., infiltration, which is a renewable resource, albeit unintendedly so. Ventilation can be defined as the process by which clean air is provided to a space. It is needed to meet the metabolic requirements of occupants and to dilute and remove pollutants emitted within a space. Usually, ventilation air must be conditioned by heating or cooling in order to maintain thermal comfort and, hence, becomes an energy liability. Indeed, ventilation energy requirements can exceed 50% of the conditioning load in some spaces [12]. Thus, excessive or uncontrolled ventilation can be a major contributor to energy costs and global pollution. Therefore, in terms of cost, energy, and pollution, efficient ventilation is essential. On the other hand, inadequate ventilation can cause comfort or health problems for the occupants.

Good indoor air quality may be defined as air, which is free of pollutants that cause irritation, discomfort or ill health to occupants [12]. Since long time is spent inside buildings, considerable effort has focused on developing methods to achieve an optimum indoor environment. An almost limitless number of pollutants may be present in a space, of which many are at virtually immeasurably low concentrations and have largely unknown toxicological effects [13]. The task of identifying and assessing the risk of individual pollutants has become a major research activity [e.g., 12, 13, and 14]. In reality, a perfectly pollutants-free environment is unlikely to be attained. Some pollutants can be tolerated at low concentrations, while irritation and odour often provide an early warning of deteriorating conditions. Health related air quality standards are typically based on risk assessment and are either specified in terms of maximum-permitted concentrations or a maximum allowed dose. Higher concentrations of pollutants are normally permitted for short-term exposure than are allowed for long-term exposure [14].

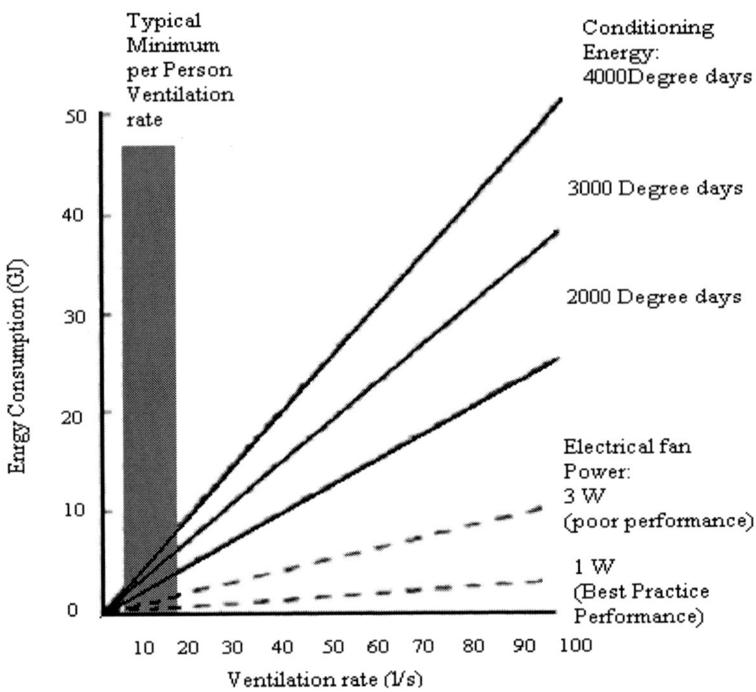

Figure 22. Energy impact of ventilation.

Ventilation is essential for securing a good indoor air quality, but, as explained earlier, can have a dominating influence on energy consumption in buildings. Air quality problems are more likely to occur if air supply is restricted. Probably a ventilation rate averaging 7 l/s.p represents a minimum acceptable rate for normal odour and comfort requirements in office type buildings [15]. Diminishing returns are likely to be experienced at rates significantly above 10 l/s.p [15]. If air quality problems still persist, the cause is likely to be poor outdoor air quality (e.g., the entrainment of outdoor traffic fumes), poor air distribution or the excessive release of avoidable pollutants into space. However, the energy efficiency of ventilation can be improved by introducing exhaust air heat recovery, ground pre-heating, demand controlled ventilation, displacement ventilation and passive cooling [16]. In each case, a very careful analysis is necessary to ensure that the anticipated savings are actually achievable. Also, it is essential to differentiate between avoidable and unavoidable pollutant emissions. Achieving energy efficiency and optimum Indoor Air Quality (IAQ) depends on minimising the emission of avoidable pollutants.

10.6. Achieving Energy Efficient Ventilation

A recent review of the International Energy Agency [17] concludes that the thermal insulation characteristics of buildings improve; ventilation and air movement is expected to become the dominant heating and cooling loss mechanism in buildings of the next century. Poor air quality in buildings sometimes manifests itself; refer to a range of symptoms that an occupant can experience while present in the building. Typical symptoms include lethargy, headaches, lack of concentration, runny nose, dry throat and eye and skin irritation [18]. Other examples of sick building syndrome have been associated with the presence of specific pollutants, such as outdoor fumes entering through air intakes [19]. Improved ventilation is one way of tackling the problem. Many standards are being introduced to ensure the adequacy and efficiency of ventilation. However, to be effective, standards need to address the minimum requirements of comfort, operational performance, air tightness, provision for maintenance and durability [20].

10.7. Ventilation of Spaces in Humid Climate

The design of windows in modern buildings in a warm, humid climate can be influenced either by their use to provide physiological and psychological comfort via providing air and daylight to interior spaces or by using them to provide aesthetically appealing fenestration. Most spaces in modern buildings are not adequately ventilated and it is recommended that effort should be directed towards the use of windows to achieve physiological comfort. Evaluation of public housing has focused on four main aspects: economics, social and physical factors; and residents satisfaction. However, information about the physiological characteristics of spaces in a warm humid climate will aid the design of appropriate spaces with respect to the development and adequate choice of building materials and appropriate use of suitable passive energy. In this light there is a need to examine residents' satisfaction with respect to these physiological issues. Proper ventilation in a space is a primary factor in determining human health, comfort and well being of the occupants. At present, getting a proper naturally ventilated space seems to be a difficult task. This is partly due to the specific environmental problems of high temperature, high humidity, low wind velocity, and variable wind direction- usually attributed to the warm humid climate, on the one hand, and the difficulty of articulating the design constraints of security, privacy and the desire of users for large spaces on the other hand.

10.8. Air movement in Buildings

Natural ventilation is now considered to be one of the requirements for a low energy building designs. Until about three decades ago the majority of office buildings were naturally ventilated. With the availability of inexpensive fossil energy and the tendency to provide better indoor environmental control, there has been a vast increase in the use of air-conditioning in new and refurbished buildings. However, recent scientific evidence on the impact of refrigerants and air-conditioning systems on the environment has promoted the more conscious building designers to give serious considerations to natural ventilation in non-domestic buildings. Two major difficulties that a designer has to resolve are the questions of airflow control and room air movement in the space. Because of the problem of scaling and the difficulty of representing natural ventilation in laboratory, most of the methods used for predicting the air movement in mechanically ventilated buildings are not very suitable for naturally ventilated spaces. However, computational fluid dynamics (CFD) is now becoming increasingly used for the design of both mechanical and natural ventilation systems. Since a CFD solution is based on the fundamental flow and energy equations, the technique is equally applicable to a naturally ventilated space as well as a mechanically ventilated one, providing that a realistic representation of the boundary conditions are made in the solution.

11. Environmental Challenge

The Montreal Protocol and its amendments (London 1990, Copenhagen 1992 and Vienna 1994) have given rise to the drawing up of regulation on the phase out the substances named Ozone Depleting Substances (ODS), including CFCs such as R12, R11 and HCFCs such as R22. Alternatives to CFCs include:

- Hydro Fluoro Carbons (HFCs) such as R134a and many mixtures.
- Naturally occurring refrigerants such as NH_3, hydrocarbons, water, air CO_2. Some of them are flammable and/or toxic or should be limited to relatively small niches.
- For other technologies such as absorption, solid sorption, etc., there are also some very interesting niches. However, refrigeration as a whole is not likely to shift there very soon because of the many other challenges it has to face.

The United Nations Framework Convention on Climate Change came into effect on March 21, 1994 and the Kyoto Protocol on December 10, 1997. Global warming is mostly caused by the rising percentage of various substances in the atmosphere, particularly CO_2,

CH_4 and many others including CFCs, Hydro Chloro Fluoro Carbons (HCFCs) and HFCs. Refrigeration is involved because of:

- Refrigerants themselves, if they are released into the atmosphere, this is the 'direct effect'.
- CO_2 produced by energy consumption of the system throughout its lifespan, this is the 'indirect effect'.

The Fifth Framework Programme (FP5) is designed to ensure that European research efforts are translated more effectively into practical and visible results. The programme includes four thematic and three horizontal programmes.

Thematic programmes are:

- Quality of life and management of living resources.
- Users-friendly information society.
- Competitive and sustainable growth.
- Energy, environment and sustainable development.

Horizontal programmes are:

- Confirming the international role of community research.
- Promotion of innovation and encouragement of participation.
- Improving human research potential and the socio-economic knowledge base.

Research in agro-food industry, including refrigeration, has an important role, particularly under the key action ''Food, Nutrition and Health''.

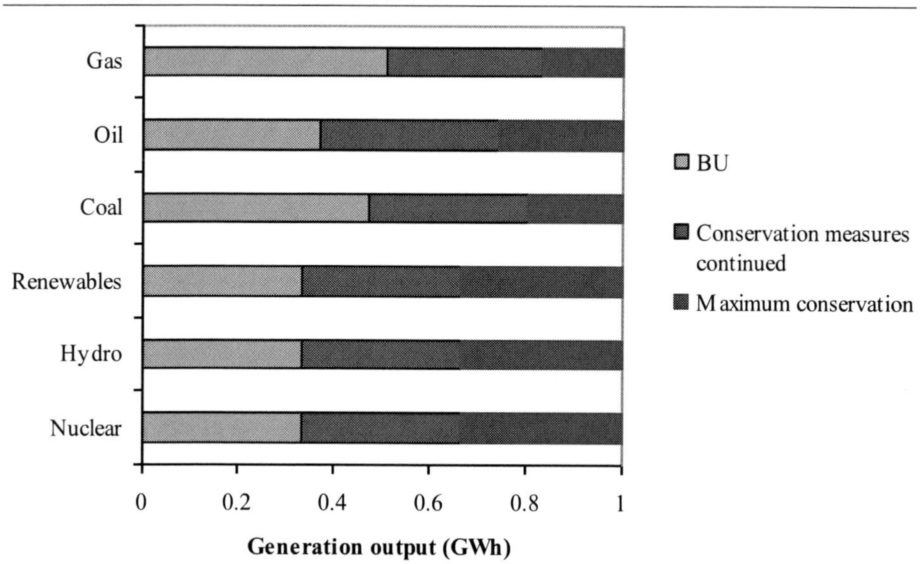

Figure 23. Energy conservation measures.

11.1. Natural Disasters

The following is a briefed guide for communities and emergency operations team to develop and maintain a viable disaster management and recovery plan. In addition, it explains in details the concept of disaster management and various steps from planning to prepare against any disaster. It further highlights the role of government agencies and local authorities at the time of disaster as well as before and after it. It also discusses long and short-term goals for mitigation, planning and recovery from disaster along with initiatives that the US government has taken in recent years. It also, has a special focus on management of energy infrastructure during the time of disaster and presents a checklist for emergency response and recovery (Figure 23).

Every year, tornadoes, hurricanes and other natural disasters injure and kill thousands of people and damage billions of dollars worth of property in the United States. Most of the times, it is almost impossible to prevent the occurrence of these disasters and their damages. However, it is possible to reduce their impact by adopting suitable disaster management strategy. Disaster management is a systematic approach towards preparing for disaster before it happens and includes disaster response - emergency evacuation, quarantine, mass econtamination - as well as supporting and ebuilding society after natural disasters have occurred. Efficient disaster management relies on thorough integration of emergency plans at all levels of government and non-government involvement.

Disaster preparedness, emergency management and post disaster recovery is highly dependent on economic and social conditions local to the disaster. However, the basics steps for disaster management remain same in all scenarios. Preparedness is the first step to counter disaster, which involves developing plan of action and it, includes communication, chain of command development, proper maintenance and training of emergency services and development of emergency warning systems along with emergency shelters and evacuation plans. Next step is response, which includes mobilisation of the necessary emergency services such as fire fighters, police, and ambulance that may be supported by a number of secondary emergency services, such as specialist rescue teams. Recovery from disaster involves restoration of the affected area including destroyed property, re-employment and redevelopment of essential infrastructure. Mitigation efforts attempt at preventing hazards from developing into disasters or reducing the impact of disasters and it focuses on long-term measures for reducing or eliminating future risks.

Conclusions

Thermal comfort is an important aspect of human life. Buildings where people work require more light than buildings where people live. In buildings where people live the energy is used for maintaining both the temperature and lighting. Hence, natural ventilation is rapidly becoming a significant part in the design strategy for non-domestic buildings because of its potential to reduce the environmental impact of building operation, due to lower energy demand for cooling. A traditional, naturally ventilated building can readily provide a high ventilation rate. On the other hand, the mechanical ventilation systems are very expensive. However, a comprehensive ecological concept can be developed to achieve a reduction of electrical and heating energy consumption, optimise natural air condition and ventilation, improve the use of daylight and choose environmentally adequate building materials.

The following are concluded:

- Promoting innovation and efficient use of applicable renewable energy technologies.
- Identifying the most feasible and cost effective applications of renewable energy resources suitable for use.
- Highlighting the local, regional and global environmental benefits of renewable energy applications.
- Ensuring the renewable energy takes its proper place in the sustainable developments, supply and use of energy for greatest benefit of all, taking due account of research requirements, energy efficiency, conservation and cost criteria.
- Ensuring the financing of and institutional support for economic renewable energy projects.
- Encouraging education, research and training in renewable energy technology in different regions in the world.

12.1. Utilisation of Renewable Energy

Developing and implementing the use of renewable energy sources.

12.1.1. Policy and Environment

- Efficiency, conservation, and policies.
- Renewable energy availability.
- Local and environmental concerns.
- Technology transfer.
- Financing requirements.
- Educational initiatives, and legislative benchmarks.

12.1.2. Solar Electrical Technology

- PV technology manufacture, testing and certification.
- PV Stand-alone systems and components.
- PV for utility rural development and grid connection.
- PV Markets and commercialisation, financing schemes and national programmes.

12.1.3. Solar Thermal Technology

- Solar radiation prediction and analysis.
- Solar cooling, heating and rural applications.
- Solar thermal applications for power generation.
- Collector technology developments.

12.1.4. Solar and Low Energy Architecture

- External Environment.
- Building, landscape design and urban communities and comfort productivity and health issues.
- Sustainable policy and social issues.
- Building simulation, material and design.
- Building refurbishment and integration of renewable energy; case studies.
- Internal environment.
- Thermal, ventilation and air movement.
- Lighting operation and control.

12.1.5. Wind Energy Technology and Applications

- Small, micro-generation and hybrid systems.
- Machines and wind farms.
- Offshore wind power.
- Wind resources and environmental issues.
- Connection and integration.
- National and regional programmes.
- Economic and institutional issues.

12.1.6. Biomass Conversion

- Heat and electricity generation.
- Energy crops and residues.
- Liquid fuels.
- Socio-economics, case studies, and environmental impacts.
- Gasification processes.

12.1.7. Fuel Cells and Hydrogen Technology

- Fuel cells technology advances electricity generation.
- Fuel cells technology advances for transportation.
- Hydrogen production for fuel processing and systems.
- Proven commercialisation reports including economic and policy issues.

12.1.8. Marine/Ocean Energy

- Wave and tidal energy resources and their characterisation.
- Device modelling, testing and development.
- Device hydrodynamics, structural integrity and environmental analysis.
- Environmental impact assessment and standards.
- Balance of system - power take-off, sensors, controls and grid integration.

- Legislation, policy, finance and markets.
- Socio-economic assessment, education and training.

REFERENCES

[1] Brain, G., and Mark, S. Garbage in, energy out: landfill gas opportunities for CHP projects. *Cogeneration and On-Site Power* 8 (5): 37-45. 2007.

[2] Federal Ministry for the Environment. Nature conservation and nuclear safety. Waste incineration- a potential danger. September 2005.

[3] WRAP. Environmental benefits of recycling: an international review of life cycle comparisons for key materials in the UK recycling sector. 2006

[4] Kyoto Protocol (CDM). 2007.http://cdm.unfccc.int/methodologies/Tools/EB35_repan 12_ Tool_grid_emission

[5] GHG Protocol. 2008. http://www.ghgprotocol.org/files/electricity

[6] Lund, J.W., Freeston, D.H., and Boyd, T.L. Direct application of geothermal energy: 2005 Worldwide Review. *Geothermics* 2005; 34: 691-727.

[7] Rawlings, R.H.D. Technical Note TN 18/99 – Ground Source Heat Pumps: A Technology Review. Bracknell. The Building Services Research and Information Association. 1999.

[8] Oxburgh, E.R. Geothermal energy. *Aspects of Energy Conversion*. p. 385-403. 1975.

[9] Erich, P. Forward facing up to climate change, in Global Climate Change and Life on Earth. R.C. Wyman (ed), Chapman and Hall, London. 1991.

[10] ASHRAE. Energy efficient design of new building except new low-rise residential buildings. BSRIASHRAE proposed standards 90-2P-1993, alternative GA. American Society of Heating, Refrigerating, and Air Conditioning Engineers Inc., USA. 1993.

[11] Molla, M. Air pollutants and its probable transmutation in the ionosphere. *Renewable Energy* 10 (2/3): 327-329. 1997.

[12] Jeremy, L. The energy crisis, global warming and the role of renewables. *Renewable Energy World* 2005; 8 (2).

[13] Omer, A. Low energy building materials: an overview. In: Proceedings of the Environment 2010: Situation and Perspectives for the European Union. p. 16-21. Porto: Portugal. 6-10 May 2003.

[14] UNEP. Handbook for the international treaties for the protection of the ozone layer. United Nations Environment Programme. Nairobi: Kenya. 2003.

[15] Viktor, D. Ventilation concepts for sustainable buildings. In: Proceedings of the World Renewable Energy Congress VII, p. 551, Cologne: Germany. 29 June – 5 July 2002.

[16] Lam, J.C. Shading effects due to nearby buildings and energy implications. *Energy Conservation and Management* 2000; 47 (7): 647-59.

[17] Raja, J., Nichol, F., and McCartney, K. Natural ventilated buildings use of controls for changing indoor climate. In: Proceedings of the 5th World Renewable Energy Congress V, p. 391-394. Florence: Italy. 20-25 September 1998.

[18] Limb, M.J. Air intake positioning to avoid contamination of ventilation. AIVC. 1995.

[19] Miller, G. Resource conservation and management. Wadsworth Publishers. California: USA, p.51-62. 1990.
[20] Abdeen M. Omer, Principle of low energy building design: Heating, ventilation and air conditioning, Cooling India: India's Premier Magazine on the Cooling Industry, Vol.5, No.4, p.26-46, Mumbai, India, October-November 2009.

In: Ventilation: Types, Standards and Problems
Editors: V. A. Romano and A. S. Duval, pp. 49-90
ISBN 978-1-61324-424-1

Chapter 2

INDUSTRIAL VENTILATION SYSTEMS

Vittorio Betta,[1,*] Furio Cascetta,[2,†] Marilena Musto,[1,‡] and Giuseppe Rotondo[1,#]
[1]University of Naples Federico II, DETEC, Italy
[2]Second University of Naples, DIAM

1. FUNDAMENTALS OF VENTILATION AND OF FLUID-DYNAMICS

1.1. Introduction

Ventilation refers to the continuous supply and removal of air with respect to a space. In industry, this is done to control chemical and some physical hazards, as well as to maintain conditions of temperature and relative humidity, which are compatible with human habitation and industrial operation.

To move air requires creating a pressure difference between two points. Air will then move from the region of higher to the region of lower pressure, at a rate that depends on the magnitude of the pressure difference and on the impedance to airflow offered by ducts, objects and friction.

The pressure difference is created with a fan or blower (mechanical ventilation), or by inducing a density difference through differential heating (natural ventilation). The latter mechanism is the source of wind outdoors but is also used in workplaces where heating sources exist, as in metal refining, melting and casting.

The desired performance of the ventilation system will impose a requirements on the magnitude of the pressure difference to be created by the fan or heat source.

* E-mail: vittorio.betta@email.it.
† E-mail: furio.cascetta@unina2.it.
‡ E-mail: marilena.musto@unina.it.
# E-mail: giuseppe.rotondo@unina.it.

In many systems, the moving air must capture the airborne contaminant by entraining the surrounding air and conducting it in the desired direction. This leads to requirements for air velocity (speed and direction) and volumetric flow rate.

Once captured, the contaminant must be carried away from the inhabited space to a suitable collector or disposal point. This is generally done by forcing the air to flow through ducts where impedance to airflow must be overcome.

In some cases, the principal goal of the system is to dilute the contaminant to safe levels by supplying a sufficient volumetric flow rate. Again, there is often a requirement that the direction of air motion be controlled to achieve the desired dilution and worker protection.

The design or selection of an industrial ventilation system is strictly related to the laws of fluid flow (fluid-mechanics or fluid-dynamics). In industrial ventilation systems, the involved fluid can be air or a mixture (air and contaminants, which are in solid, liquid or gaseous state). Knowledge of these laws is fundamental for design engineers, and a thorough understanding of basic theory is essential to successful solution of ventilation problem.

This paragraph reviews briefly the fundamentals of fluid flow that have significant application to air movement in industrial ventilation systems.

1.2. Properties of Standard Air

The standard air is an ideal gas at the following thermodynamic conditions: 1.013 bar, 20 °C (dry bulb temperature) and zero water content (dry air).

This is the standard definition of air in current use for industrial and engineering ventilation systems. In the above standard air conditions, the reference density (ρ) is considered equal to 1.20 kg/m^3.

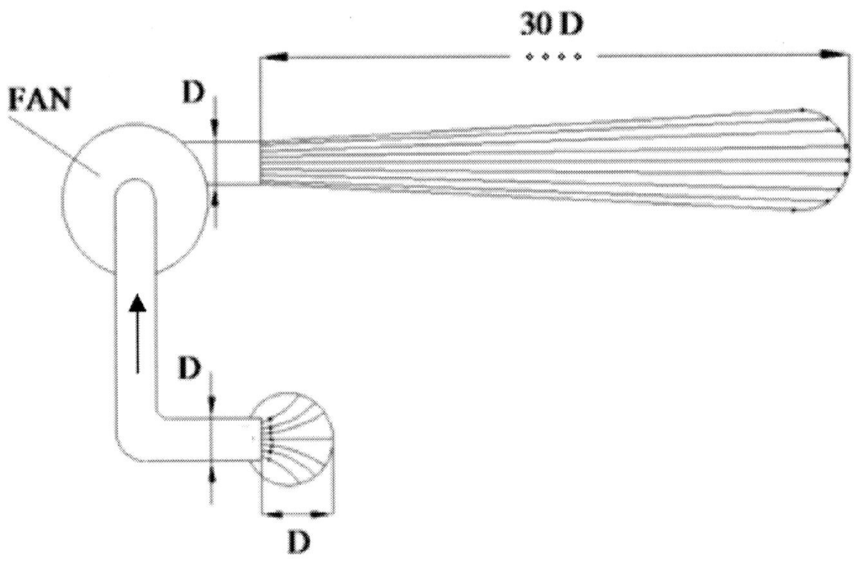

Figure 1.1. Qualitative comparison: Jet flow vs. suction flow.

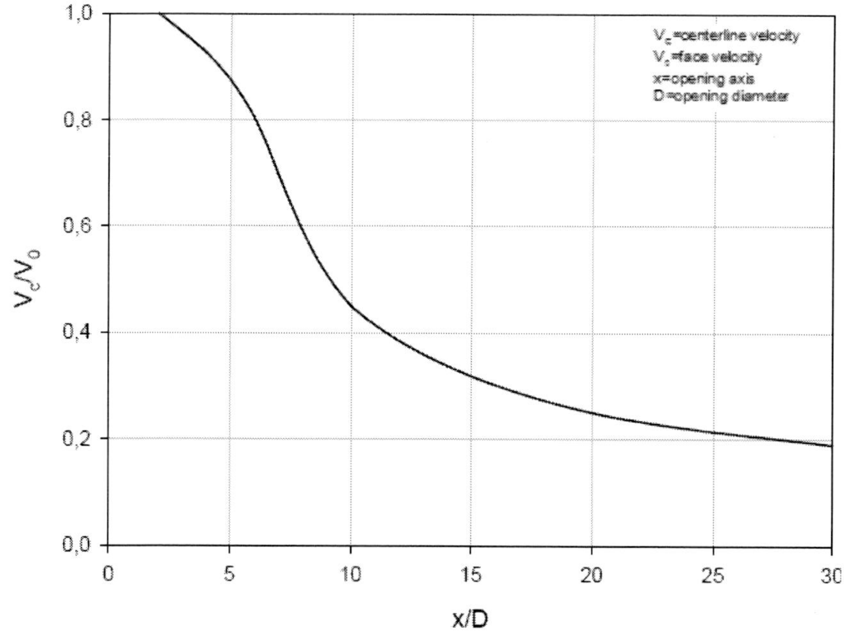

Figure 1.2. Centerline (axial) velocity decay of the in a push flow.

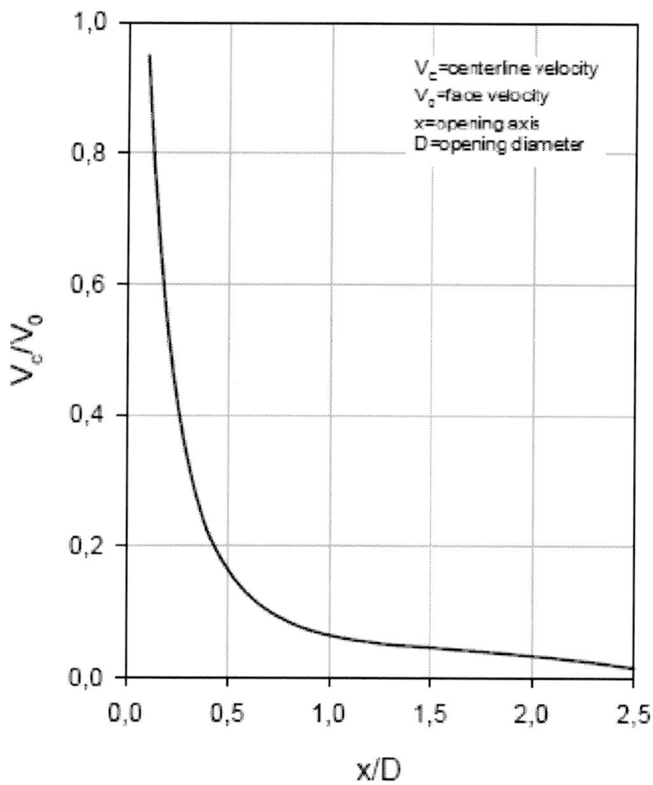

Figure 1.3. Centerline (axial) velocity decay of the in a pull flow.

1.3. Push or Jet Flow

Push flow (or jet flow) occurs in jets when blowing is taking place.

The push flow maintains its initial flow velocity for a long distance. For example (Figures 1.1 and 1.2), the push flow maintains the velocity equal to 20% at a distance of 30 D (D=diameter of the opening). In industrial ventilation, the rule of push flow is as a controller and/or a carrier.

1.4. Pull or Suction Flow

In the pull flow suction or inhaling of the air occurs.

The pull flow velocity decreases steeply to only 7% at a distance of 1 D (Figures 1.1 and 1.3). In industrial ventilation, the rule of pull flow is as a receiver.

1.5. Bernoulli's Equation and Consequences

Bernoulli's theorem is of considerable practical importance to the designer of exhaust systems. The Bernoulli's theorem expresses the basic principles of energy conservation.

Before introducing the Bernoulli's equation, it is important to provide the following basic fluid-dynamics definitions.

Static pressure (p_s), in fluid dynamics, is the pressure at a nominated point of a fluid. Many authors use the term "static pressure" in place of "pressure" to avoid ambiguity. The concepts of "static pressure," "total pressure" and" dynamic pressure" arise from Bernoulli's equation and are significant in the study of the fluid flows. *Static pressure* is identical to *pressure* and can be identified for every point in a fluid flow field.

Dynamic pressure (p_d), is proportional to the kinetic energy of the air stream; in particular, it is equal to half the product of the fluid density and the squared velocity (1/2 ρv^2). From Bernoulli's equation for an incompressible flow, the p_d is equal to the difference between the total pressure (p_t) and the static pressure (p_s):

$$p_d = p_t - p_s = 1/2\ \rho v^2.$$

Dynamic pressure (p_d) is also called *Velocity pressure* (v_p): in incompressible flow is $p_d=v_p$.

Total (impact) pressure (p_t) is the pressure that would arise if a moving fluid were brought to rest without change of entropy (adiabatic process or isocaloric process). It is also defined as the algebraic sum of static and dynamic pressures, and it can be positive or negative with respect to atmospheric pressure:

$$p_d = p_t - p_s$$

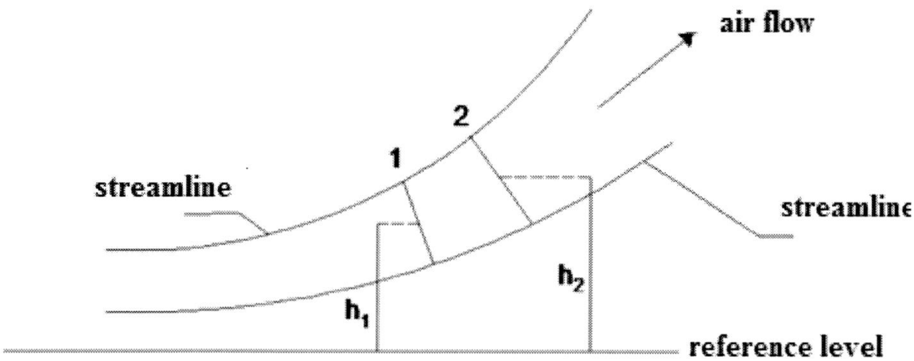

Figure 1.4. Flow through a stream tube.

Incompressible flow is the flow in which changes in the fluid density due to the motion can be neglected. Although air is a compressible fluid, in many cases, for low velocity of air, the flow can be treated as incompressible.

Compressible flow is the flow in which the motion produced appreciable changes in the fluid density.

The Bernoulli's equation in the most common form is:

$$v dv + g dh + \frac{dp}{\rho} = 0 \tag{1}$$

on integration along a stream tube (Figure 1.4), this equation becomes:

$$\frac{v^2}{2} + gh + \int \frac{dp}{\rho} = \text{const} \tag{2}$$

where

ρ is the fluid density,

g is the acceleration due to the gravity (also called gravitational acceleration),

h is the level.

In incompressible flow, the total pressure (p_t) is equal to the sum of the static pressure (p_s) and the velocity pressure ($1/2\ \rho v^2$):

$$p_t = p_s + \frac{1}{2}\rho v^2 \tag{3}$$

In compressible flow, the Bernoulli's equation allows the following relationship:

$$p_t = p_s \left[1 + \frac{\gamma - 1}{2} M^2 \right]^{\frac{\gamma}{\gamma - 1}} \tag{4}$$

where

γ is the ratio of the specific heat of the air at constant pressure (c_p) to that at constant volume (c_v):

$$\gamma = \frac{c_p}{c_v} \tag{5}$$

M is the Mach number defined as:

$$M = \frac{v}{a} = \frac{v}{\sqrt{\frac{\gamma p_s}{\rho}}} \tag{6}$$

where a is the speed of sound in air.

Applying the Taylor series to the equation (4), it becomes:

$$p_t = p_s + \frac{1}{2}\rho v^2\left[1 + \frac{M^2}{4} + \frac{M^4}{40} + \ldots\right] \tag{7}$$

For example, at ordinary temperature, when Mach number (M) assumes value less than 0.2 (corresponding to air velocity v less than 62 m/s and velocity sound in air about 341 m/s), the second term in the square bracket ($M^2/4$) in the expansion of equation (7) becomes equal to 0.01 so it is negligible compared to 1 and, therefore, in this case, the flow motion can be considered incompressible. In most cases of ventilation systems, the air velocity value is much smaller than 62 m/s; just in a few cases, higher air velocity values required an approach of compressible flow.

The mass flow rate (kg/s) is defined as:

$$Q_m = \rho \cdot \overline{v} \cdot A \tag{8}$$

where $\overline{v}$ is the average velocity in the cross-section area A.

The volumetric flow rate (m^3/s) is defined as:

$$Q_V = \overline{v} \cdot A \tag{9}$$

1.6. Flow Regimes and Velocity Distributions

Reynolds number is a characteristic dimensionless group that expresses the ratio between the inertial forces to the viscous forces. For internal flow in a circular conduit, the Reynolds number can be referred to the characteristic length (the diameter D of the pipe):

$$\mathrm{Re}_D = \frac{\text{inertial forces}}{\text{viscous forces}} = \frac{\rho v D}{\mu} = \frac{vD}{\nu} \tag{10}$$

where μ is the dynamic viscosity, ρ is the air density and v is the air velocity and ν is the kinetic viscosity ($\nu=\mu/\rho$).

For conduits with different cross-section (rectangular, square, etc…), the characteristics length is the hydraulic diameter D_H (defined as 4 A/P, where A is the cross-section area, and P is its perimeter):

$$\mathrm{Re}_{D_H} = \frac{\rho v D_H}{\mu} \tag{11}$$

Low Reynolds numbers ($\mathrm{Re}_D < 2300$) imply that the inertial forces are negligible with respect to the viscous forces: in this case, the flow regime is called *laminar*. Laminar flow regime may be thought of as consisting of thin layers all parallel to each other. Laminar flow is common only in cases in which the fluid is moving slowly, and its viscosity is relatively high.

For very high Reynolds numbers (Re_D □ 2300), the viscous forces are negligible with respect to the inertial forces: in this case, the flow regime is called *turbulent*. Turbulent flow regime represents the case in which the fluid undergoes irregular fluctuations or mixing. The speed of the fluid at a point is continuously undergoing changes in magnitude and direction, which results in swirling and eddying as the bulk of the fluid moves in a specific direction.

The transitional regime is the situation in which the velocity distribution is changing from laminar to turbulent flow.

In Figure 1.5 are represented the velocity distributions (velocity profiles) in a circular pipe for laminar and for turbulent flows.

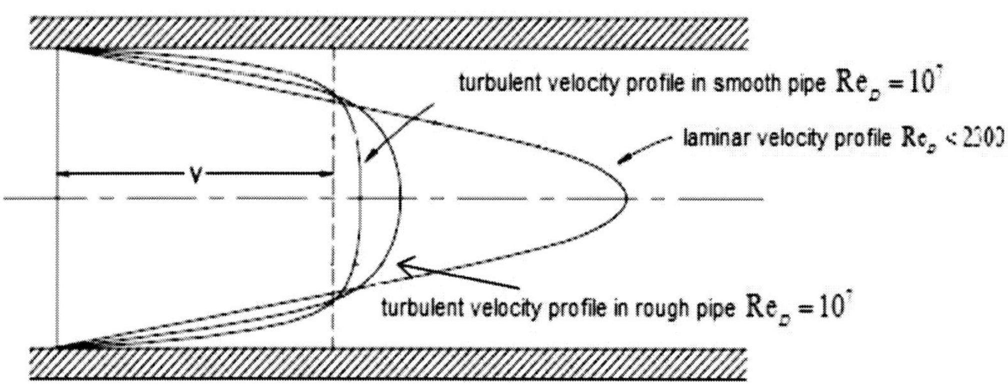

Figure 1.5. Velocity profiles in different flow regimes.

1.7. Navier-Stokes Equations

The Navier-Stokes equations describe how the velocity, pressure, temperature and density of a moving fluid are related. They are a set of coupled differential equations and consist of a time-dependent continuity equation for conservation, three time-dependent conservation of momentum equations (x, y, z momentum) and a time-dependent conservation of energy equation. There are four independent variables in the problem: the x, y, and z spatial coordinates of a domain and the time t. There are six dependent variables: the pressure (p), density (ρ, temperature (T)—which are contained in the energy equation through the total energy Et—and three components of the velocity vector [u(x), v(y), w(z)]. All of the dependent variables are functions of all the four independent variables. The five differential equations are, therefore, partial differential equations:

continuity equation:

$$\frac{\partial \rho}{\partial t}+\frac{\partial(\rho u)}{\partial x}+\frac{\partial(\rho v)}{\partial y}+\frac{\partial(\rho w)}{\partial z}=0 \tag{12}$$

x-momentum equation:

$$\frac{\partial}{\partial t}(\rho u)+\frac{\partial(\rho u^2)}{\partial x}+\frac{\partial(\rho uv)}{\partial y}+\frac{\partial(\rho uw)}{\partial z}=-\frac{\partial p}{\partial x}+\frac{1}{\mathrm{Re}_r}\left[\frac{\partial \tau_{xx}}{\partial x}+\frac{\partial \tau_{xy}}{\partial y}+\frac{\partial \tau_{xz}}{\partial z}\right] \tag{13}$$

y-momentum equation:

$$\frac{\partial(\rho v)}{\partial t}+\frac{\partial(\rho uv)}{\partial x}+\frac{\partial(\rho v^2)}{\partial y}+\frac{\partial(\rho vw)}{\partial z}=-\frac{\partial p}{\partial y}+\frac{1}{\mathrm{Re}_r}\left[\frac{\partial \tau_{xy}}{\partial x}+\frac{\partial \tau_{yy}}{\partial y}+\frac{\partial \tau_{yz}}{\partial z}\right] \tag{14}$$

z-momentum equation:

$$\frac{\partial(E_T)}{\partial t}+\frac{\partial(\rho uw)}{\partial x}+\frac{\partial(\rho vw)}{\partial y}+\frac{\partial(\rho w^2)}{\partial z}=-\frac{\partial p}{\partial z}+\frac{1}{\mathrm{Re}_r}\left[\frac{\partial \tau_{xz}}{\partial x}+\frac{\partial \tau_{yz}}{\partial y}+\frac{\partial \tau_{zz}}{\partial z}\right] \tag{15}$$

conservation equation:

$$\frac{\partial(E_T)}{\partial t}+\frac{\partial(uE_T)}{\partial x}+\frac{\partial(vE_T)}{\partial y}+\frac{\partial(wE_T)}{\partial z}=-\frac{\partial(up)}{\partial x}-\frac{\partial(vp)}{\partial y}-\frac{\partial(wp)}{\partial z}-\frac{1}{\mathrm{Re}_r\,\mathrm{Pr}_r}\left[\frac{\partial q_x}{\partial x}+\frac{\partial q_y}{\partial y}+\frac{\partial q_z}{\partial z}\right]$$

$$+\frac{1}{\mathrm{Re}_r}\left[\frac{\partial}{\partial x}(u\tau_{xx}+v\tau_{xy}+w\tau_{xz})+\frac{\partial}{\partial y}(u\tau_{xy}+v\tau_{yy}+w\tau_{yz})+\frac{\partial}{\partial z}(u\tau_{xz}+v\tau_{yz}+w\tau_{zz})\right] \tag{16}$$

In theory, the Navier-Stokes equations can be solved for a given flow problem (with known initial and boundary conditions) by analytical methods. In practice, these equations are too difficult to solve analytically, since they are non-linear and no general exact solution is obtainable. Just in a few simple cases, the Navier-Stokes equations can be resolved (allowing the exact solutions):

1) steady and unsteady parallel flows: between parallel walls, both fixed; between parallel walls, one fixed and one moving in its own plane, with or without a positive or negative pressure gradient (*Couette flow*); within a closed conduit of any arbitrary cross-section; on one side of a rotating plate; between concentric rotating cylinders.
2) steady flows: between rotating plates; between convergent and divergent plates; plane and axially symmetric flows normally to a plate (stagnation flows); flow in a laminar jet of circular cross-section.
3) unsteady flows: adjacent to a suddenly accelerated plane wall; adjacent to a plate oscillating in its own plane; dissipation of an initially irrotational vortex tube by viscous shear.

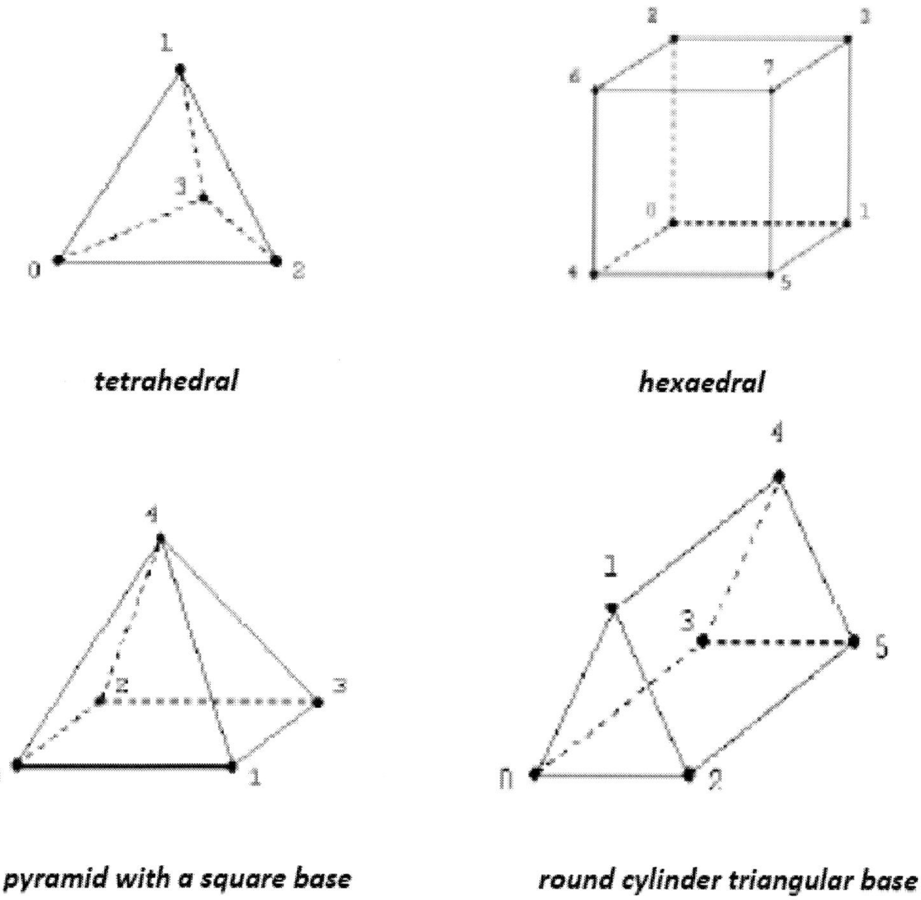

Figure 1.6. Different cell shapes.

Recently, high-speed computers have been used to solve approximations to the equations using a variety of numerical techniques, such as finite difference, finite volume, finite element, and spectral methods. This area of study is called Computational Fluid Dynamics or CFD.

The numerical CFD approach is becoming increasingly used; it has enabled, regardless of the complexity of a geometrical area and boundary spatial and time conditions, relatively easy and, at the same time, very precise prediction of even very complex fields of velocity and temperature. This particular advantage of the CFD approach has made it an almost ideal method for designing and optimization of the ventilation.

Commercial numerical CFD simulators are now available on the market (for example, popular CFD codes are Fluent, FDS), mainly based on finite volume approach. In such software, the user typically follows three steps:

1) geometry discretization of the fluid-dynamic domain under investigation: a discretization scheme involves the breaking-down of the geometry into smaller parts (elements, volumes, points, etc.). The resulting discretized geometry, which is now a group of several smaller computational regions, is termed as a grid. The approximate formulations of the governing equations obtained through the different discretization schemes are solved on the discretized geometry or grid. The grid has more entities and can be of different shapes as shown in the Figure 1.6.
2) setting the initial and boundary conditions. The user can define:
   - different mechanisms and models (e.g., turbulence models, moving/deforming mesh);
   - incorporate different physics (e.g., phase change, real-gas formulations)
   - adjust the numerical parameters (e.g., face-approximation techniques, under-relaxation factors);
   - convergence criteria (e.g., residuals accepted, lift and drag coefficients monitors);
   - time-step for calculations (not valid for steady-state analysis).
3) processing and data analysis. Important concept for CFD analysis is that of boundary conditions. Since a finite flow domain is specified, physical conditions are required on the boundaries of the flow domain. Each of the dependent variable equations requires meaningful values at the boundary of the calculation domain in order for the calculations to generate meaningful values throughout the domain. These values are known as boundary conditions and can be specified in a number of ways. The simulation generally starts from an initial solution and uses an iterative process to reach a final flow field solution. CFD problems in general are non-linear, and the solution techniques use an iterative process to successively improve a solution, until *convergence* has been achieved. The main indicator for convergence is the point at which the finite difference equations for all cells are satisfied by the current values of the appropriate dependent variables. However, it can be very useful in many cases to monitor the variation of the dependent variables at specific locations throughout the calculation domain in order to observe the point at which they stabilize. It may be acceptable, in many cases, to terminate a calculation as soon as the dependent variables have achieved adequate stability but before the residuals have arrived at

their defined termination values. Post-Processing involves extracting the desired flow properties (temperature, velocity, pressure, etc...) from the computed flow field.

The goodness and the reliability of the numerical simulation results are strictly related to the knowledge, the ability and the experience of the user in CFD problems.

2. Classification and Applications of Industrial Ventilation Systems

2.1. Introduction

Industrial ventilation is a method of controlling worker exposure to airborne toxic chemicals or flammable vapors by exhausting contaminated air away from the work area and replacing it with clean air.

There are three types of workplace ventilation:

1. *indoor air-quality ventilation*, used primarily to provide fresh, heated or cooled air to buildings, as part of the heating, ventilating and air-conditioning systems; this ventilation type is mainly used in civil applications, where contaminants are not particularly toxic or dangerous;
2. *dilution ventilation,* which dilutes contaminated air in a whole building or room by blowing-in clean air and exhausting the same dirty air; such approach in large industrial environment is so expansive since it implies movement and conditioning of relevant air quantity; this ventilation type is not applicable in industrial hazardous areas, where chemical, dust, vapor, fume, explosive contaminants may be present;
3. *local exhaust ventilation,* which captures contaminate emissions at source (or very near) and exhausts them outside; such approach is preferable in widespread industrial applications and allows both a good contaminant control and energy saving. In this paragraph are detailed the local exhaust ventilation systems in the main industrial applications.

Local exhaust ventilation systems typically consist of the following basic elements:

- hood or entry point of the system to capture pollutants and/or contaminants;
- duct system to transport polluted air;
- air-cleaning device to remove captured pollutants from the airstream;
- air-moving device (fan or air ejector);
- exhaust stack, which discharges system air to the atmosphere;
- return duct system to return cleaned air back to the plant.

In this section, the attention is focused on the different types and performance of the local exhaust hoods.

The classification of local exhaust hoods can be carried out for the following three different approaches:

a) by *contaminant source type*: knowledge of the process is essential before a hood can be designed. The type, size and geometry of the hood depend on the pollution sources. There are three types of pollution sources:

a. 1) buoyant (heat) sources, cause contaminants to move in a buoyant plumes;

a. 2) non buoyant (diffusion) sources, create contaminant diffusion in all directions due to the concentration gradient;

a. 3) dynamic sources, create contaminant movement with an air jet .

b) by *hood type*:

b. 1) *enclosing hoods* provide better and economical contaminant control, because their exhaust rates and the effects of room air currents are minimal. Although enclosing hoods provide the best control, they are often not feasible because they would interfere with the work being done by the employee. In those cases, a capture exhaust hood can only be located near the source of the contaminant. However, the distance between the face of the hood and source must be short to effectively capture the contaminant. Adding a flange to the edges of the capturing hood provides more efficient capture of contaminants.

b. 2) *exterior hoods* can be used if access requirements make it necessary to leave all or part of the process open; exterior hoods should be located so that the contaminant is drawn away from the operator's breathing zone. Canopy hoods should be used where the operator must bend over a tank or process.

c) by *system mobility:* local exhaust systems with non-enclosing hoods can be

c. 1) stationary (i.e., having a fixed hood position);

c. 2) moveable/portable/turnable: used when the process equipment must be accessed for repair, loading and unloading.

c. 3) Built-in (into the process equipment).

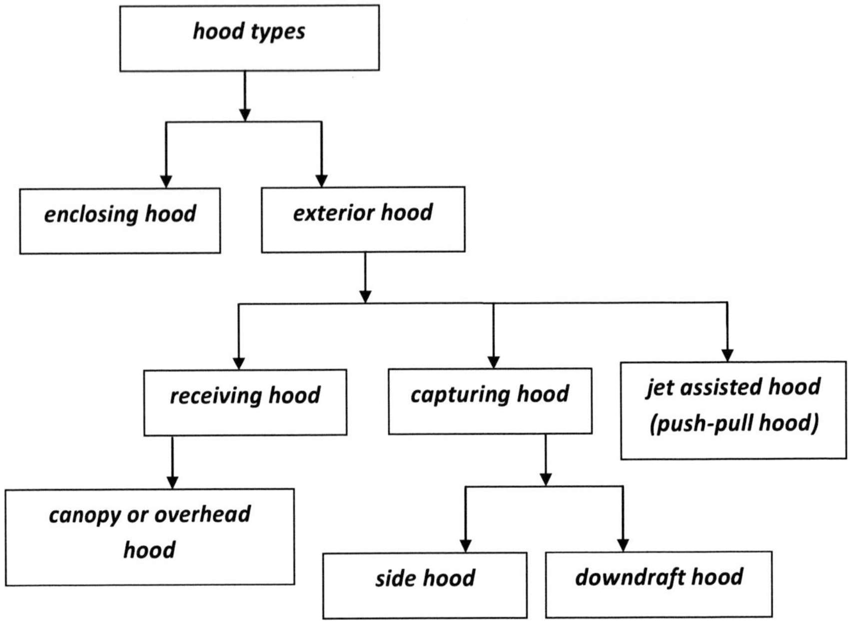

Figure 2.1. Hood types classification.

2.2. Hood Types and Applications

Hoods are designed to confine or capture the contaminant at their sources. The air velocity at the hood opening and inside the hood must be sufficient to capture and carry the air contaminants. The hood should enclose the source of contaminant as much as possible or be placed as close to the source as possible.

Hoods may be of a wide range of physical configurations. Typically, hoods can be grouped in two general categories (Figure 2.1):

2.2.1. Enclosing Hoods

Enclosing hoods are those that completely or partially enclose the process of contaminant generation point (Figure 2.2). Applications of enclosing hoods are available in laboratory glove box or similar type of enclosure where only minimal openings exist. Complete enclosures are usual for abrasive blasting equipment, tumbling mills, mixer, bucket elevators, storage bins, glove boxes for clean rooms or toxic material operation, vibrating and rotary screens, automatic buffing, foundry shakeout, automatic painting and dipping, crucible and heat-treating furnaces. Enclosing hoods surround the contaminating source more or less completely. Hence, for this type of hood, it is important to have an efficient "draught" in order to allow the dusts and/or fumes to escape outside the volume of the hood, thus confining the process. Where possible, this type of hood is preferred because obviously the better the hood is able to "wrap round" the contaminating source, the smaller the air flow rate required for controlling the atmosphere. Obviously, enclosing hoods are not useable in all those situations where the operator is required to take an active part in the production process. Typical applications of enclosing hoods are those concerning the transport and conveying of solid particulate (Figure 2.2, right side) (for example, carbon dust by means of belt conveyor).

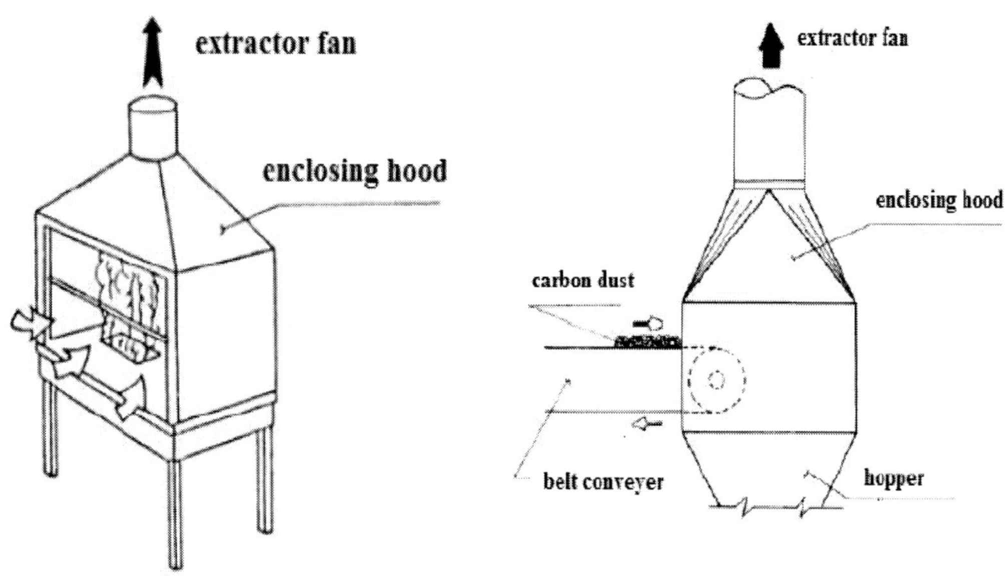

Figure 2.2. Examples of enclosing hoods.

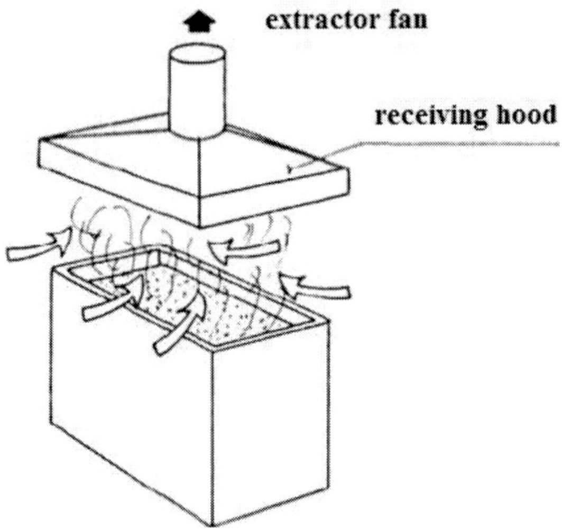

Figure 2.3. Canopy hood type.

2.2.2. Exterior Hoods

Exterior hoods are located adjacent to an emission source without enclosing it, where the contaminant is gas/vapor/fine particulate and is not emitted with any significant velocity. To achieve higher capturing effectiveness, the exhaust should be located as close to the source as possible. If the contaminant contains large particulates that are emitted with a significant velocity, the hood should be located in the emission zone. The exhaust opening can be mainly of two geometries: rectangular (square) or circular. In addition, the exhaust opening may be free or flanged. The exterior hoods can be classified in three principal hood types: receiving, capturing and jet assisted.

Receiving Hoods

Receiving hoods are configured so as to "spontaneously" capture the contaminating elements released by a particular process. Normally, these hoods are used either in "hot" processes or in those situations where the contaminating particles, of medium-large size, are released at a sufficient speed to reach the hood. Vapors/fumes produced by a hot source naturally tend to rise (natural convective motions) towards the hood, which, as its name suggests, receives them and disposes of them.

Usual application of a *canopy* hood is over hot processes where natural convection produces a vertical thrust to the contaminant (thermal convective plume). It should not be used in the case that contaminants precipitate by gravity on the way to hood's opening (powder, organic solvent). Canopy hoods, in general, require low airflow rate since the contaminant moves naturally toward the hood.

A typical exterior canopy hood is illustrated in Figure 2.3; the distance of the canopy hood above the process may range from 1 m (low canopy hood) up to 3 m (high canopy hood). High canopy hoods require higher volumetric flow rates with respect to the low canopy hood.

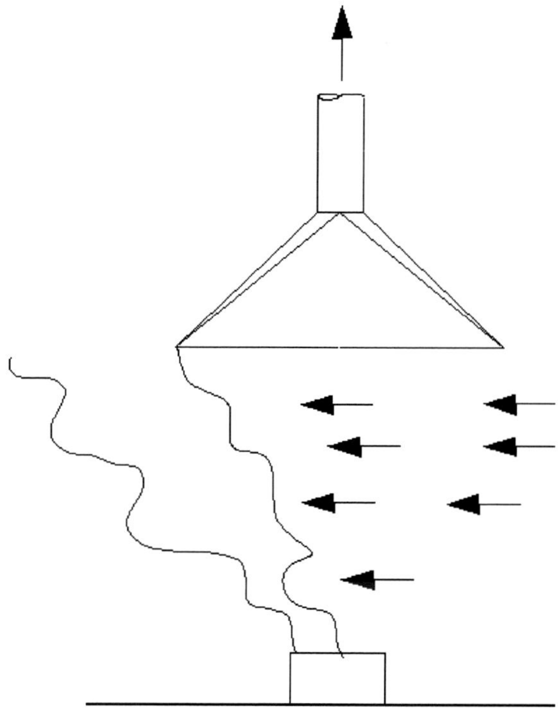

Figure 2.4. Cross-draft effect on thermal plume under high hood.

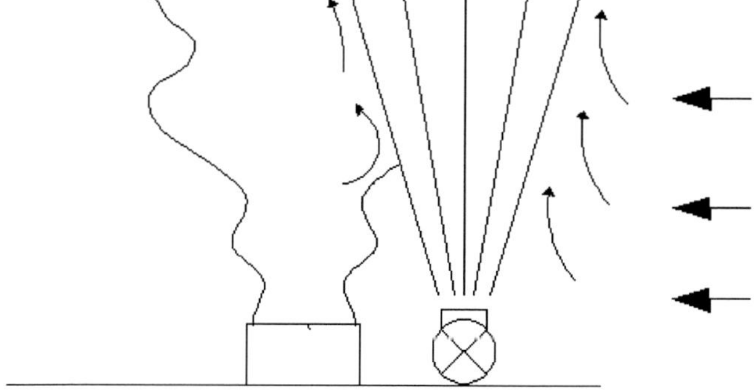

Figure 2.5. High-hood protection with single-side air curtain.

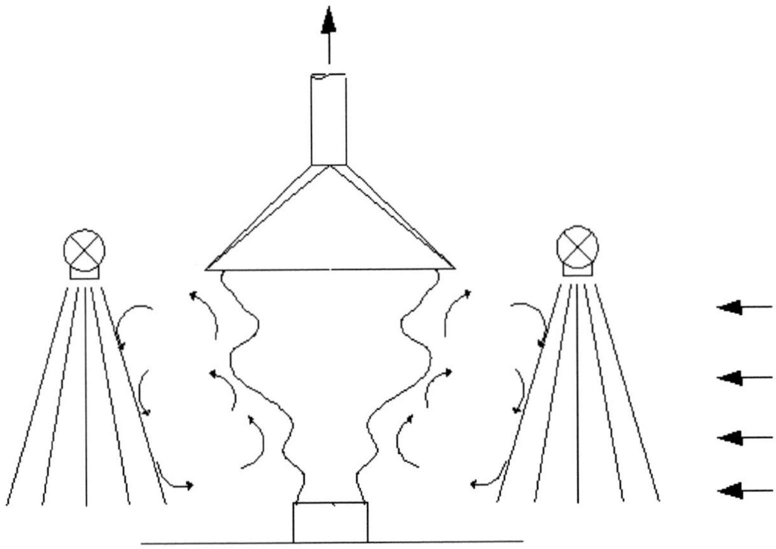

Figure 2.6. High-hood protection with double-side air curtain.

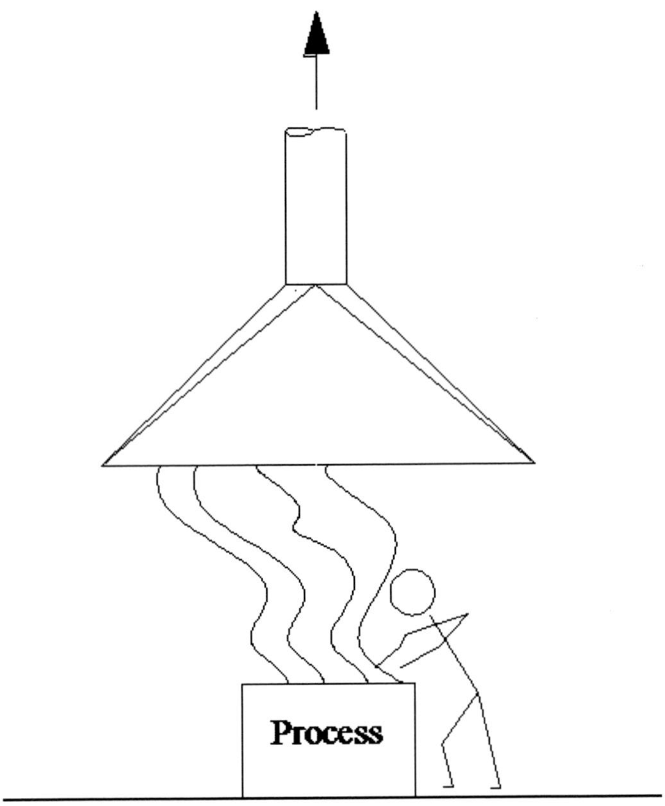

Figure 2.7. Bad interference between airflow and operator's breathing zone.

High canopy hood can be influenced by the presence of cross-draft flows, which deviate the thermal plume outside the suction zone of the hood (Figure 2.4). To overcome this problem, it is possible to protect the high canopy hood operation by means of air jet (air curtain), both with single side (Figure 2.5) and with double side (Figure 2.6). Canopy hoods should not be used when there is an interference between the velocity field generated and the operator's breathing zone (Figure 2.7).

Where contaminants are not moved by natural convective motion, the receiving hoods are located along the preferential path of the emitted contaminant.

In function of the contaminant emission typology (nature of particles, their velocity and mass), the hoods can be positioned at the top or at the side. It should also be stated that in receiving hoods, the fan (or extractor) performs the sole function of disposing of the volumes of air contained in the hood and already not that of capturing the contaminating particles at the end of the hood. Moreover, use of the receiving hoods is recommended when the contaminating process to be controlled takes place in devices of considerable size, i.e., to discourage, for reasons of cost-effectiveness, the use of enclosing hoods, because the roll forming of the metal sheets around the contaminating sources would be too difficult. The main limit of receiving hoods is that of not being able to capture the particles of smaller size (less than approx. 30 μm) because the latter do not possess sufficient quantity of motion to reach the hoods with the main flow. It is well known that particles of small size are the ones most harmful for person's health, because if inhaled, they are deposited in the lungs, unlike the coarser particles, which are captured in the nose or at the most deposited in the upper respiratory ways. This is why such hoods cannot be used in processes requiring the presence of an operator who, for reasons stated above, would inevitably be in a position to inhale the released contaminates in his work zone. A typical example of application of receiving hoods, is the one regarding certain types of chip-cutting machining operations (e.g., grinding, sawing, polishing), i.e., where the contaminating particles produced possess sufficient quantity of motion and preferential paths (Figure 2.8).

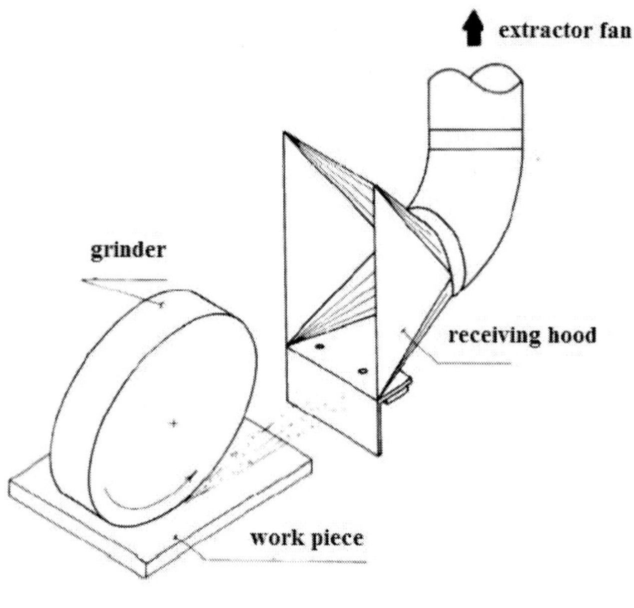

Figure 2.8. Receiving hood type.

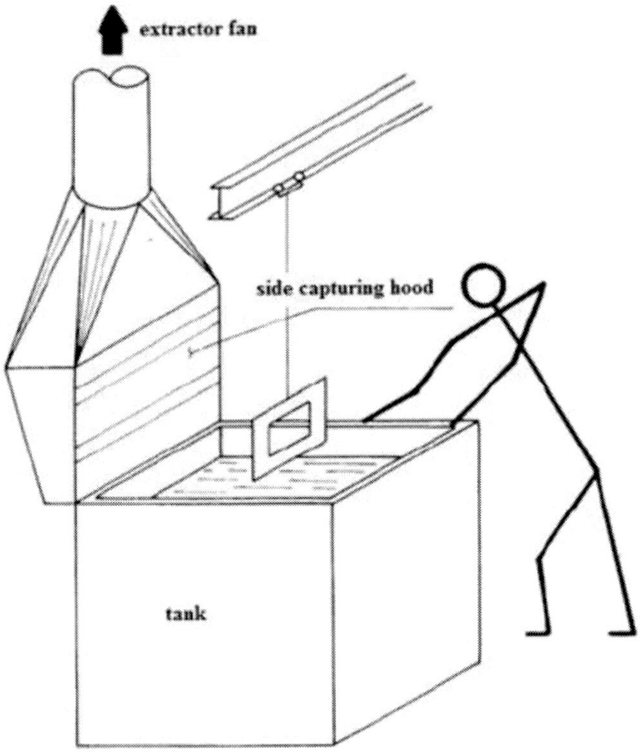

Figure 2.9. Lateral/side multislot capturing hood.

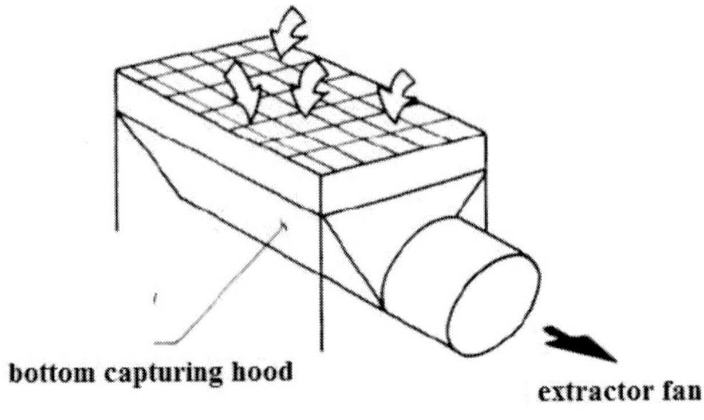

Figure 2.10. Bottom/downdraft capturing hood.

Capturing Hood

For processes requiring operator attendance, it is advisable to adopt capturing hoods, since they can be located with a minimal interference with operators' working zones.

In capturing hoods, the contaminant, necessarily of small size (dia. < 30 μm), is drawn towards the hood, even from relatively modest distances (max. 60-70 cm) by means of a directional airflow, such as to obtain the appropriate capture velocity at the required distance. However, compared to the receiving hoods, the capturing hoods have greater running costs owing to the large volumes of air required to ensure an adequate capture velocity in the vicinity of the contaminating source. The capturing hoods can, in turn, be classified into side and bottom types.

The main advantage of *lateral/side capturing* hoods consists of the poor interference between the breathing zone of the operator and the velocity field generated (Figure 2.9).

A sidedraft hood requires a higher exhaust volumetric flow rate than a canopy hood.

In the *downdraft hoods,* as the name implies, exhaust of contaminant is downward, usually through a floor grille or a grille in a workbench top (Figure 2.10). It should be considered only when overhead and sidedraft are impractical. Performance is best where contaminant is released closed to the grille and where side shields can direct downflow of air over the work area and reduce short circuiting. A downdraft hood will tend to aspirate and entrain more materials than other hood types. Often, the effective area of the grille is blocked by the operation at the time when maximum contaminant release occurs, making effective control difficult.

Jet-assisted Hood (Push-Pull Hood)

Jet-assisted hoods are exterior hoods combined with compact, linear, or radial air jets. They are used: (i) to increase capturing effectiveness by transporting contaminants toward the face of the hood using supply jets, (ii) to separate contaminated zones from relatively clean zones in working spaces, (iii) to protect hoods from room air movement.

The *push-pull ventilation system* (Figure 2.11) is a more efficient alternative to lateral exhaust for open-surface vessels such as plating tanks. In a push-pull system, air is blown through one slot on the periphery of a vessel and exhausted from a slot on the opposite side. In these systems, a jet of compressed air pushes the flow of contaminating particles tangentially towards the mouth of the appropriately positioned side capture hood. The inlet to the hood must be large enough to accommodate the jet and the room air entrained by the jet.

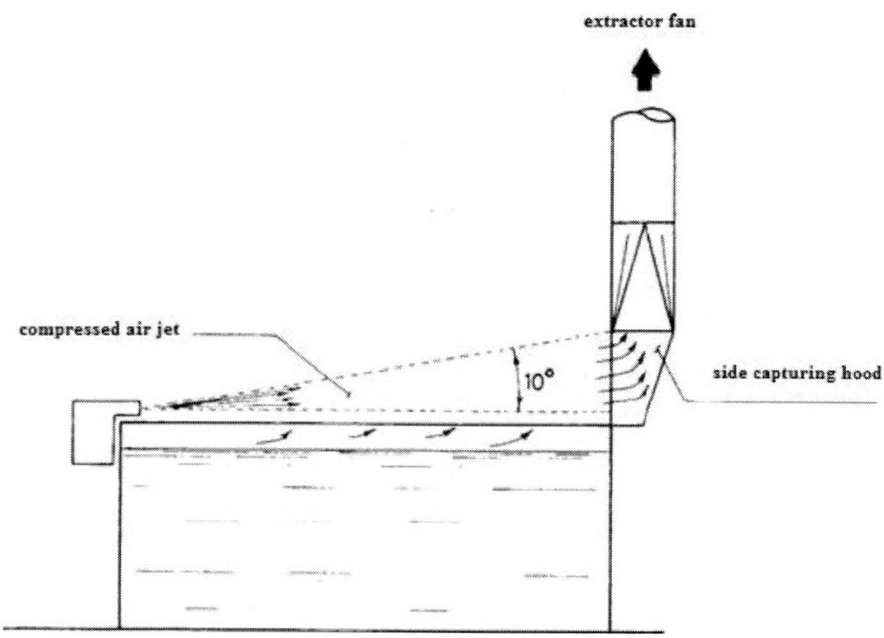

Figure 2.11. Push-pull ventilation system.

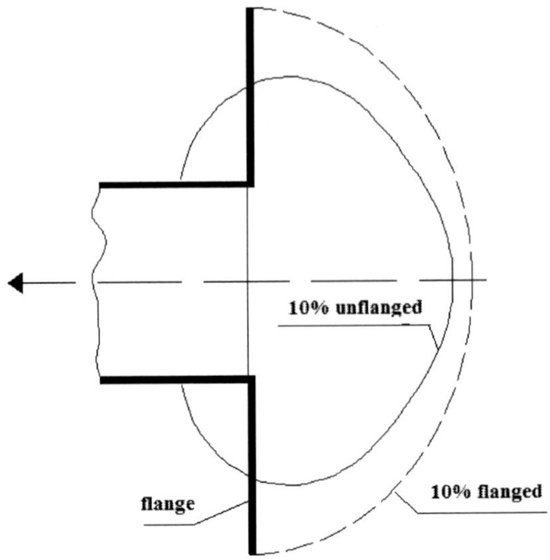

Figure 2.1.2 Velocity contour zone.

Table 2.1. Centerline air velocities induced by external hoods

Schematic	Equation	Applicable range	Reference
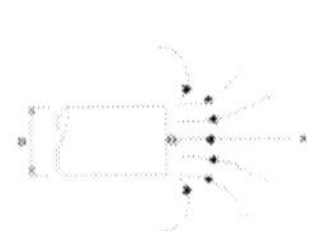	$\frac{V_x}{V_0} = \left(1+10\frac{x^2}{A}\right)^{-1}$	$x \leq 1.7\sqrt{A}$ $\alpha \leq 30°$	Dalla Valle (1952)
	$\frac{V_x}{V_0} = 1.1(0.07)^{x/D}$ $\frac{V_x}{V_0} = 0.1\left(\frac{x}{D}\right)^{-1.6}$	$0 \leq \frac{x}{D} \leq 0.5, C \geq D$ $0.5 \leq \frac{x}{D} \leq 1.5, C \geq D$	Garrison (1977)
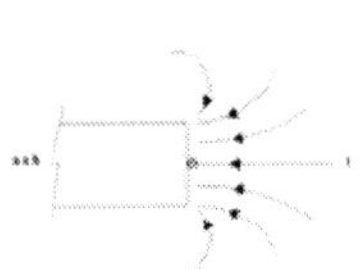	$\frac{V_x}{V_0} = 1.1\left(0.93 + 8.58\alpha_p^2\right)^{-1}$ $\alpha_p = \left(x/\sqrt{A}\right)(a/b)^{-\beta_p}$ $\beta_p = 0.2\left(x/\sqrt{A}\right)^{-1/3}$	$1 \leq \frac{a}{b} \leq 16$ $0 \leq \frac{x}{\sqrt{A}} \leq 3$ $\alpha \leq 30°$	Fletcher (1977)
	$\left[\frac{V_x}{V_0} = 1 - \frac{2}{\pi}\,\mathrm{atan}\left(\frac{2x\sqrt{x^2+a^2+b^2}}{ab}\right)^{-1}\right]$	$1 \leq \frac{a}{b} \leq 16$ $0 \leq \frac{x}{\sqrt{A}} \leq 1.6$ $\frac{C}{\sqrt{A}} \geq 1$	Tyaglo and Shepelev (1970)
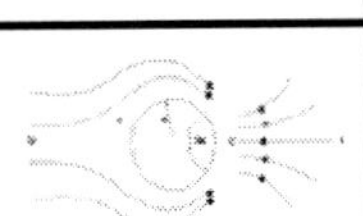	$\frac{V_{cx}}{V_0} = \frac{V_{cn}}{V_0} = \frac{2R}{\pi x}\,\mathrm{atan}\left[\frac{x+R}{x-R}\tan\left(\frac{\alpha}{2}\right)\right]$	$\lvert x \rvert \geq R$	Posokhin (1984)

2.4. Centerline Air Velocities Prediction Formulas

The freestanding hood openings may be round, rectangular or slot. In order to improve the effect of the exhaust action, usually the hood opening is flanged: the presence of the flange reduces the flow from the ineffective regions, so enlarging the velocity contour zone (Figure 2.12). In the following Table 2.1 are reported the centerline velocities for different realistic external hoods.

3. Ventilation Systems in Enclosed Vehicle Facilities

3.1. Introduction

This section covers ventilation requirements for contaminant level control and emergency smoke/temperature management in road tunnels and parking structures. Design approach for various natural and mechanical ventilation systems are also illustrated.

3.2. Road Tunnels

A road tunnel is an enclosed facility with an operating roadway for motor vehicles passing through it. All road tunnels require ventilation to remove contaminants produced during normal engine operation. Ventilation in normal conditions (i.e., without fire) may be obtained by: (i) natural means, (ii) traffic-induced piston effects, or (iii) mechanical equipment. Ventilation must also provide control of smoke and heated gases from a fire in the tunnel (emergency condition). Smoke flow control is necessary in order to realize an environment suitable for evacuation.

Ventilation in emergency conditions can be provided by: (i) natural means, (ii) taking advantage of the buoyancy of smoke and hot gases, or (iii) mechanical means.

Natural and traffic-induced ventilation systems are adequate for relatively short tunnels (up to 300 m) and for those with low traffic volume or density. Long (more than 700 m) and heavily travelled tunnels should have mechanical ventilation systems. Tunnel length ranging between 350 and 650 m require a specific approach (engineering analysis) to evaluate the needs of the suitable ventilation system.

Factors affecting ventilation system selection include:

- tunnel length;
- cross-section;
- grade or slope;
- surrounding environment;
- volume and typology of traffic;
- direction (one-way or unidirectional, two ways or bidirectional);
- construction cost;
- energy cost.

3.2.1. Mechanical Ventilation

In road tunnels, motor vehicles produce discharge exhaust gases, and, therefore, it is crucial to control pollutant concentrations in order to protect the health of passengers. Pollutants can be diluted by introducing appropriate external airflow rates. Obviously, this dilution should be performed assuming that outdoor air is less polluted than the inner one. It is possible, also, to use appropriate filters.

As concerns the road tunnel mechanical ventilation classification, it is possible to distinguish:

- *normal ventilation* is used during ordinary traffic operations to maintain acceptable levels of contaminants in the tunnel; main pollutants emitted from vehicles are carbon monoxide, oxides of nitrogen, particulate matter, unburned hydrocarbons, aromatics, solid lead, sulfur oxides; the ventilation system is aimed to control the concentration of pollutions dangerous for human health. Normal ventilation can be divided in normal and congested ventilation:
  - normal operation mode is where the tunnel fans are off and the recirculation of air within the tunnel is through the vehicle piston effect. The pollution concentration in the tunnel is less than the design pollution concentration limit;
  - congested mode is when traffic in the tunnel increases with a slow down in traffic movements, the amount of CO (carbon monoxide) and the temperature in the tunnel will increase. When this happens, the sensors in the tunnel will trigger the operation of the tunnel fans to dilute the effluents concentration.
- *emergency ventilation* is used to control and remove smoke and hot gases produced by a fire process; the main parameters to control are heat/temperature (in order to avoid the combustion/explosion of vehicles) and smoke (in order to avoid smoke poisoning); its principal aim is to provide an evacuation path that is sufficiently clear of smoke and hot gases and is maintained at a sufficiently low temperature to permit safe evacuation of motorists. Emergency ventilation systems may also be used to remove stagnant contaminants during adverse atmospheric conditions.
- *temporary ventilation* is used during original construction or while work is being done in a finished tunnel. It is removed after construction is complete.

The main mechanical ventilation systems can be grouped in three categories:

A) the *transverse ventilation system* aims to suck polluted air and replace it with clean air, continuously along the entire length of tunnel (Figure 3.1). This type of ventilation is used in extremely long tunnels and in tunnels with heavy traffic volume. This system is usually composed of a group of ventilation fans mounted in a chamber over the tunnel ceiling. In urban tunnels, usually the fans are equipped with silencers, in order to reduce noise generation. Figure 3.2 shows a typical cross - section of a transversal ventilation tunnel. This ventilation type is considered effective and safe in case of fire. During a fire, exhaust fans in the full transverse system should operate at the highest available capacity, while supply fans should operate at a somewhat lower capacity. This allows the stratified smoke layer (at the tunnel ceiling) to remain at that higher elevation and be extracted by the exhausted system.

B) the *longitudinal ventilation system* induces a longitudinal airflow along the roadway in which fresh air is introduced into or removed from the tunnel at a limited number of points, such as at the portal or a shaft, and can only be used in unidirectional tunnels. Longitudinal systems are sub-categorized into: (i) jet-fan-based ventilation systems and (ii) longitudinal ventilations with Saccardo nozzle.

- The jet-fan ventilation system (Figure 3.3 left side) employs a series of axial fans mounted at the ceiling level of the tunnel producing a high discharge thrust and velocity, which induces additional airflow within the tunnel. This system can only be used in unidirectional tunnels, as smoke and heat will be discharged from the exiting portal. This ventilation system represents a much easier and cheaper solution.
- Saccardo ventilation system (Figure 3.3 right side) introduced air into or removed from the tunnel at a limited number of points such as at the portal or a shaft (Bendelius, 2003). In long tunnels, exchanging vitiated air for fresh air through a shaft may be a better solution than a larger number of jet fans.

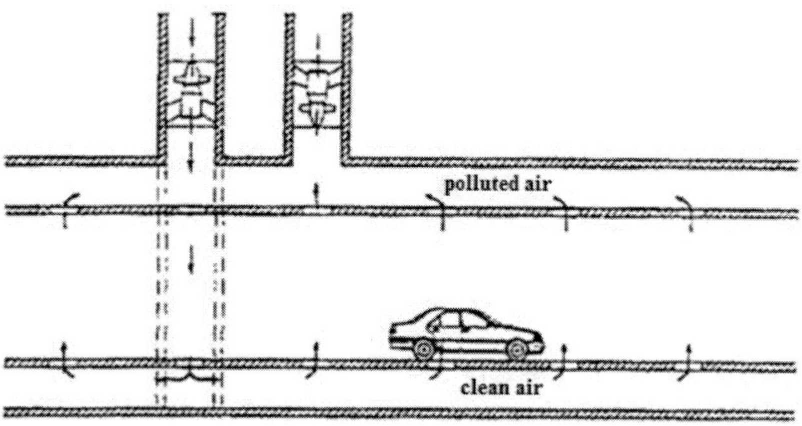

Figure 3.1. Full transverse ventilation system.

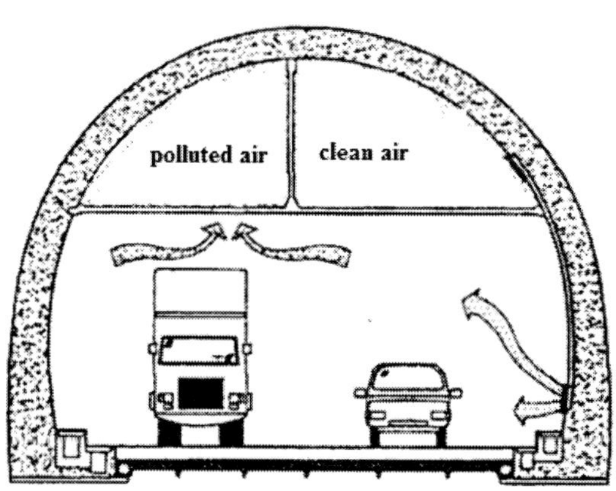

Figure 3.2. Cross-section of a full transverse ventilation tunnel.

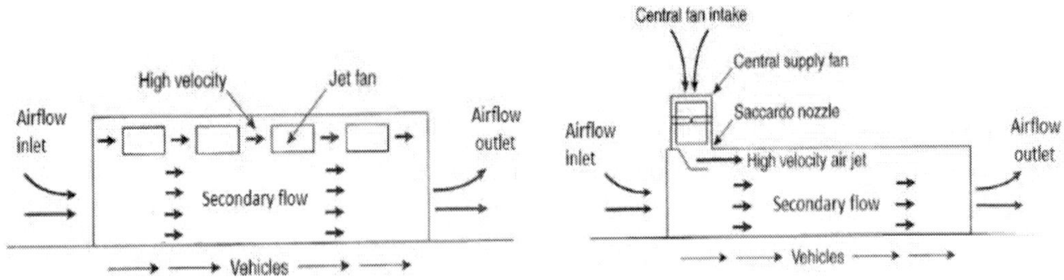

Figure 3.3. Longitudinal ventilation system.

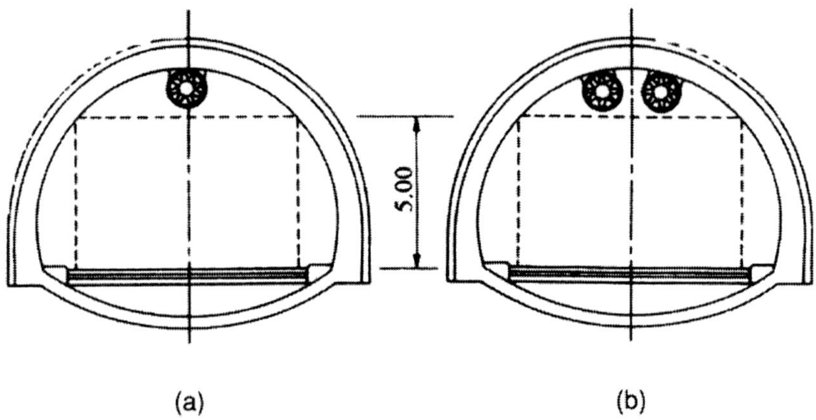

Figure 3.4. a) Typical position of a single fan; b) typical position of a pair of fans.

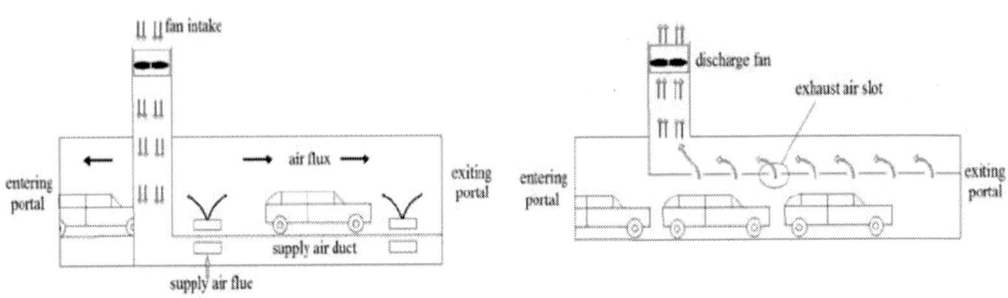

Reproduced from Bendelius (2003).

Figure 3.5. Semi-transverse ventilation system,: supply system(left side), exhaust system (right side).

Longitudinal ventilation is the most effective method of smoke control in a road tunnel with unidirectional traffic. In case of fire, a ventilation system must generate sufficient longitudinal air velocity to prevent *backlayering* of smoke (movement of smoke and hot gases against ventilation airflow in the tunnel roadway), providing protected road for the evacuation of vehicle occupants and for the action of emergency services. The air velocity to prevent backlayering over stalled or blocked motor vehicles is the minimum velocity needed for smoke control in a longitudinal ventilation system, and it is known as *critical velocity.*

In the case of a one-way tunnel (with length less than 3 km), it is preferable to move the airflow in the same direction of the vehicles (Figure 3.3), in order to improve the *piston effect* (the piston effect is the additional airflow due to the moving vehicles). Impulse fans can be installed individually or in multiple arrangements (Figures 3.4-a and 3.4-b), leaving a free height of 5 meters for passing larger vehicles.

The mutual distance between two subsequent jet fans in longitudinal ventilation is normally calculated satisfying the conditions that the inlet velocity (entering the generic jet fan) is about the mean velocity of the flow rate through the tunnel.

As concerns the maximum allowable lengths of longitudinally ventilated tunnels, the following are normally recommended:

- urban tunnel, bi-directional traffic: prohibited;
- urban tunnel uni-directional traffic: 500 to 800 m;
- non urban tunnel, bi-directional heavy traffic: 1 km;
- non urban tunnel, bi-directional light traffic: 1 to 1.5 km;
- non urban tunnel, uni-directional traffic: 5 km.

The one-way tunnel configuration assures an effective control of airflows, both in normal operation and in emergency conditions (fire).

The same level of airflow control is not obtained in the case of two-way tunnels: in such cases, the airflow moved by the jet-fans can interfere with the piston effect produced by the vehicle motion. In case of fire, the ventilation systems do not guarantee the same safety condition of the one-way tunnel configuration since vehicles can be present both upstream and downstream the fire point.

C) the *semi-transverse (hybrid) ventilation system* consists of a combination of above-said ventilation systems. This type of ventilation system is normally used in tunnels up to about 2,000 m. It is possible to design three types of semitransverse ventilation systems:

- supply semitransverse ventilation (in which the airflow is pushed inside the tunnel both along the tunnel axis and lateral air duct);
- exhaust semitransverse ventilation (in which the airflow is pulled outside the tunnel);
- supply and exhaust semitransverse ventilation (in which the airflow is both pushed and pulled through the tunnel).

3.2.2. Natural Ventilation

Naturally ventilated tunnels rely primarily on atmospheric conditions to maintain airflow and a satisfactory environment in the tunnel. The piston effect gives an additional contribution to the airflow. The chief factor affecting the tunnel environment is the pressure differential created by the differences in the elevation, the ambient temperature or wind between the two tunnel portals. Unfortunately, adverse atmospheric conditions (adverse winds at portal) may limit or block the natural venting of smoke to atmosphere stratification, or, in the case of a traffic jam, it can reduce the piston effect benefit.

For these reasons, the length of the tunnels with natural ventilation systems are limited as follows:

- urban tunnel (uni or bi-directional traffic): 300 m;

- non-urban tunnel with heavy traffic (uni or bi-directional): 500 m;
- non-urban tunnel with light traffic (uni or bi-directional): 1000 m.

Naturally ventilated tunnels can be "portal-to-portal" type or "portal–to-shaft" type. The first configuration is preferable in the uni-directional traffic, while a naturally ventilated tunnel with intermediate shaft is better suited for bi-directional traffic.

Smoke from a fire in a tunnel with only natural ventilation is driven primarily by the buoyant effects of hot gases and will tend to flow upgrade. The steeper the grade (slope), the faster the smoke will move.

3.3. Parking Garages

Automobile parking garages can be either fully enclosed or partially open:

- fully enclosed parking areas are usually underground and require mechanical ventilation systems, both in ordinary service and in case of fire;
- partially open parking garages are generally above-grade structural decks having open sides (Table 3.1), with a complete deck above; in ordinary service, the ventilation can be a combination of natural and mechanical; in case of fire, the ventilation is mainly mechanical.

Operating automobiles in parking garages presents two concerns. The more serious is emission of CO and CO_2, with its known risks; the other concern is oil and gasoline fumes, which may cause nausea and headaches and also represent potential fire hazards. Additional concerns about NOx and smoke haze from diesel engines may also require consideration. However, the ventilation rate required to dilute CO_2 to acceptable levels is usually satisfactory to control the level of other contaminants as well, provided the percentage of diesel vehicles does not exceed 20%.

Table 3.1. Comparison between International guidelines

Country	Minimal percentage open of facade			Maximum Distance between open facades	Maximum height
	Of total facade per floor	Of total facade surface	Of floor area		
Austria	25 %	50%	-	70 [m]	8 [floors]
Belgium	33 %	-	-	-	10 [m]
Germany	33 %	-	-	70 [m]	22 [m]
Finland	30 %	-	-	-	8 [floors]
Great Britain	-	-	5 %	90 [m]	15 [m]
Italy	-	60 %	-	-	-
The Netherlands	33 %*	-	2.5 %*	54 [m]	20 [m]
Norway	33 %	-	-	-	16 [m]
Switzerland	25 %	-	-	70 [m]	-
United states of America	20 %	variable	-	80 [m]	23 [m]

*) At least one of these conditions, the 2.5% must be met for each open facade.

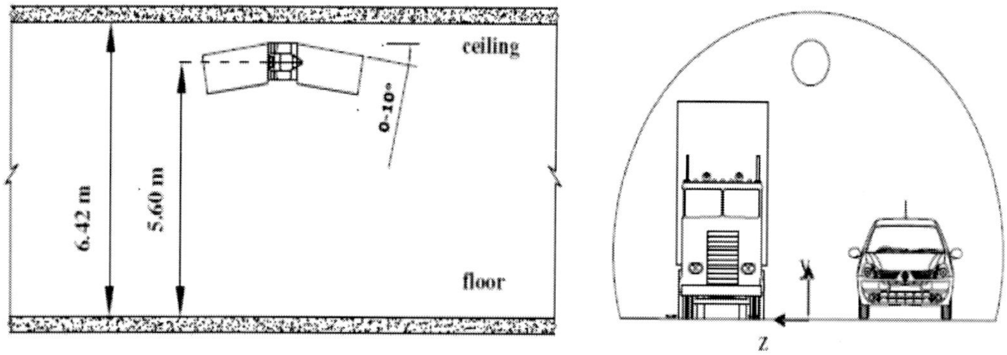

Figure 3.6. Sketch of a banana-jet installation in a longitudinal tunnel (lateral view left, section view right).

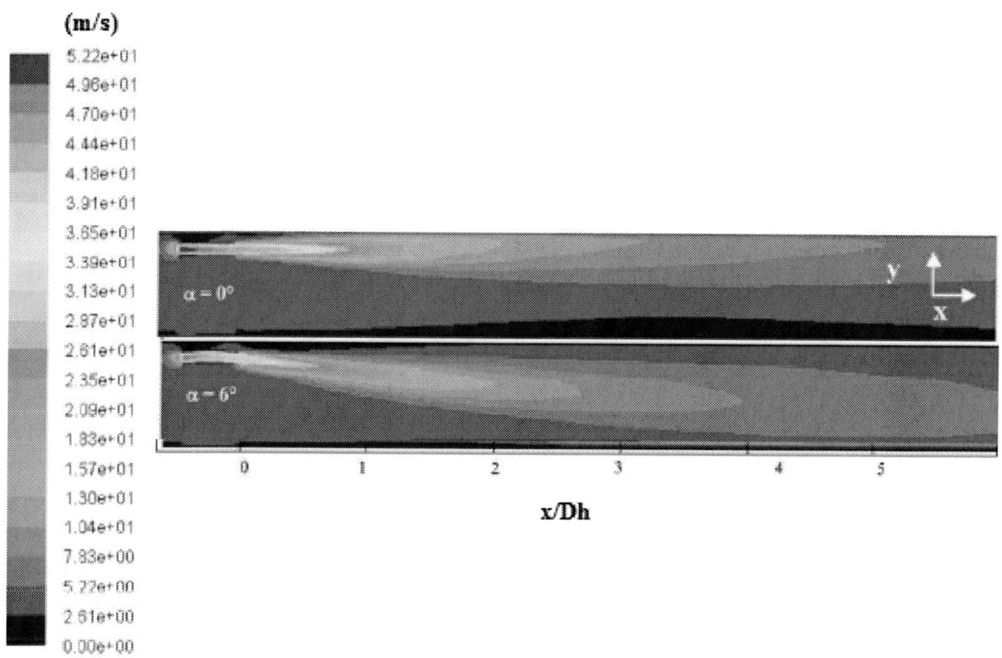

Figure 3.7. Comparison of the velocity fields generated by a traditional jet fan with α=0 ° (above) and by a banana-jet fan with α=6 ° (below) in almost free tunnel.

The design ventilation rate required for an enclosed parking facility depends chiefly on four factors:

1. acceptable level of contaminants in the parking facility;
2. number of cars in operation during peak conditions;
3. length of travel and operating time or cars in the garage;
4. emission rate or typical car under various conditions.

Table 3.2. Comparison of the performance ratio (PR)

	Free tunnel	Traffic jam tunnel
α (°)	PR	PR
0	1.00	1.00
2	1.17	1.06
4	1.31	1.06
6	1.42	0.89
8	1.36	0.68
10	1.24	0.56

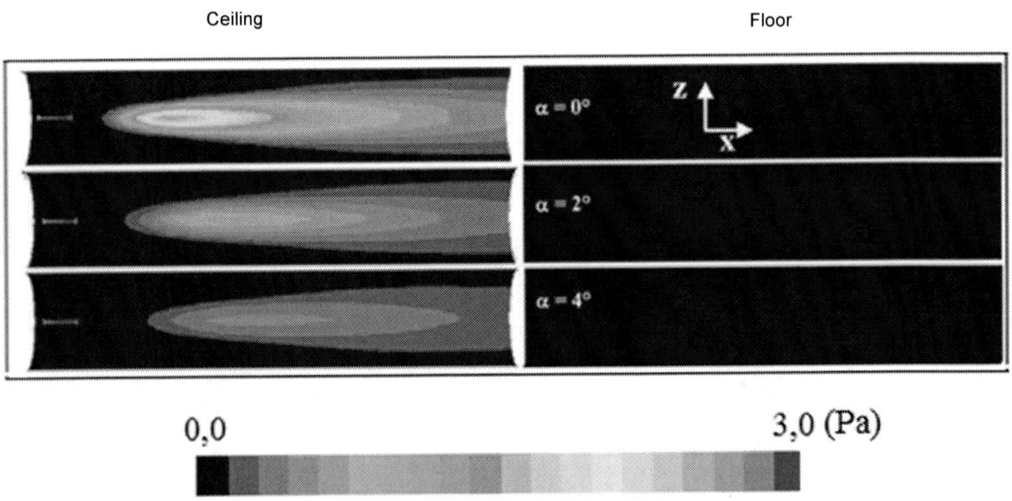

Figure 3.8. Comparison of the shear stress produced by a traditional jet fan with α=0 ° (above) and by a banana-jet fan with α=2 ° (middle) and with α=4° (below), in traffic jam configuration.

3.4. Case Studies

3.4.1. Tunnel in Ordinary Ventilation

In this section are presented the main results of a CFD simulation, aimed to evaluate the performance of ventilation of a specific jet fan called a "banana jet fan." This type of fan presents the inlet and outlet silencers inclined toward the tunnel floor (Figure 3.6) and forms a pitch angle (α) away from the ceiling. Traditional jet fans are typically with horizontal axes (pitch angle α = 0 °).

The use of a banana jet fan allows managing the tunnel ventilation system optimizing the fluid dynamic efficiency and energy consumption. In particular, for very light traffic conditions (comparable to a free tunnel condition), the optimal pitch angle configuration is α=6 °. In Figure 3.7 is shown the comparison of the simulated velocity fields generated, respectively, by a traditional jet fan (α=0 °) and by an optimized banana jet fan with α=6 °.

In the traffic jam condition, the CFD simulation results lead to define a different optimal pitch angle of the banana jet fan (2 °≤α≤4 °) as shown in Figure 3.8.

Introducing the performance ratio (PR), defined as the ratio between the thrust force for an assigned longitudinal flux in the tunnel for α=0 ° and the thrust force for an assigned longitudinal flux in the tunnel for α>0 °, it is possible to determine the optimal angle corresponding to the maximum PR value (Table 3.2).

3.4.2. Tunnel in Emergency (Fire) Ventilation

In this section are presented the main results of a further CFD simulation, aimed to evaluate the performance of ventilation of specific ventilation fan called "banana jet fan" in emergency fire conditions, both with and without traffic.

Both for free tunnel or for traffic jam tunnel, the fire was simulated setting heat flux on HVC (Heavy Good Vehicle) surface and comprehensive of radiative heat flux.

The results of CFD simulation, shown in the Figure 3.9, refer to a transversal section positioned at 50 m from burning vehicle (HVC) with traffic. It is possible to note that the optimal configuration of the banana jet fan pitch angle is between α= 2 ° and 4 °, since the temperature fields are relatively lower than for other α values, especially near the floor of the tunnel.

In the case of fire without traffic, the CFD results (Figure 3.10) show that the optimal configuration of the banana jet fan pitch angle is between α= 2° and 6°.

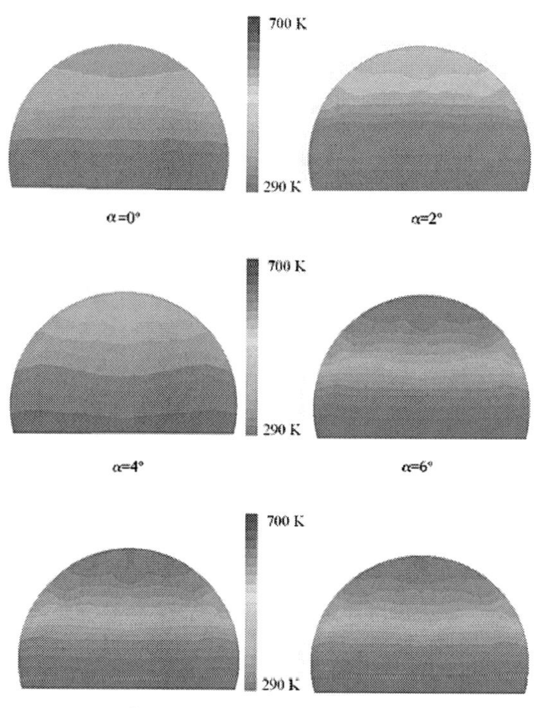

Figure 3.9. Temperature fields in a cross section at 50 m from burning HVC, in emergency fire condition with traffic.

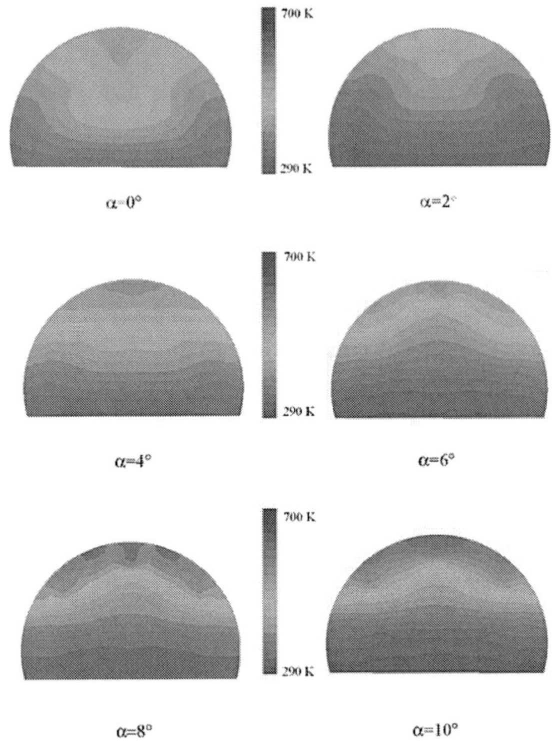

Figure 3.10. Temperature fields in a cross-section at 50 m from burning HVC, in emergency fire condition without traffic.

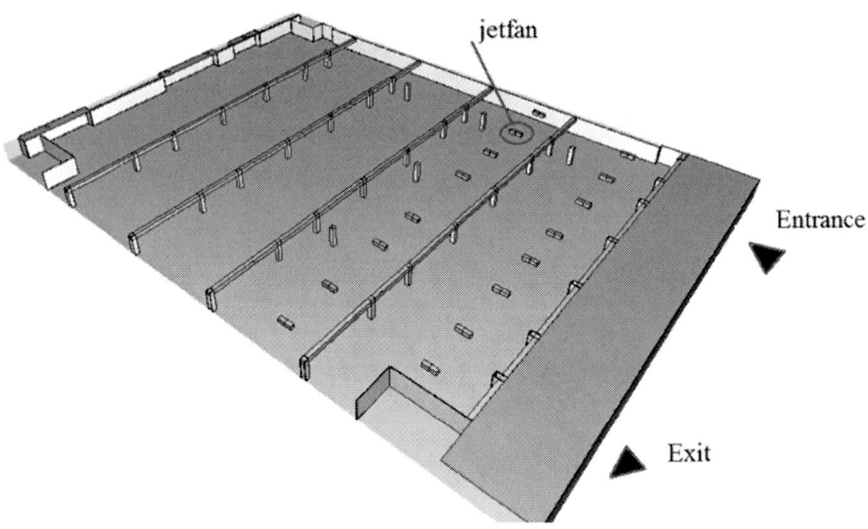

Figure 3.11. Car parking.

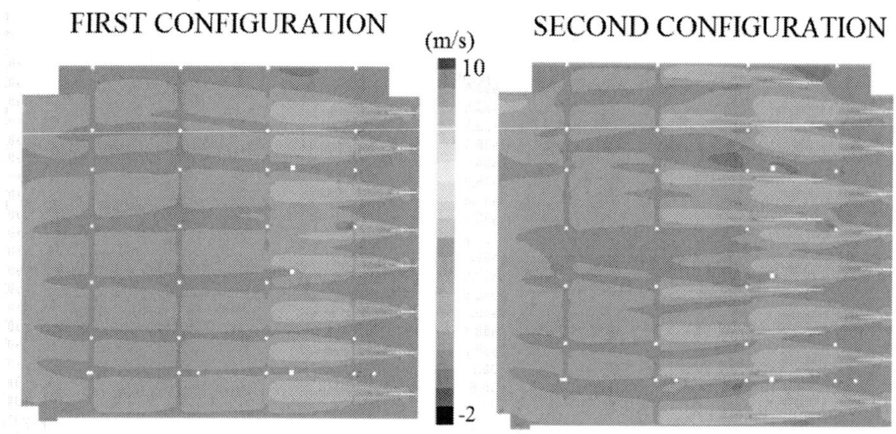

Figure 3.12. Velocity fields along the plane parallel to the floor (y = 0 m) that passes through the plane of symmetry.

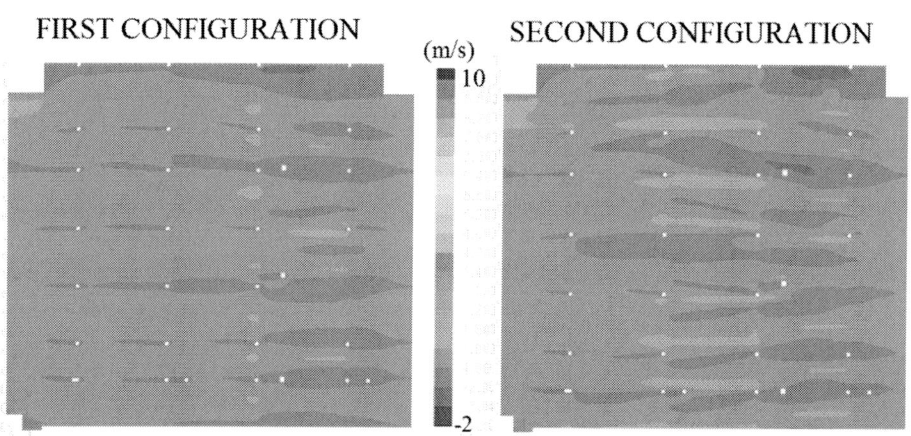

Figure 3.13. Longitudinal velocity field at y=0.5 m for both configurations.

3.4.3. Parking Ventilation System

In this section are presented the main results of a CFD simulation, aimed to evaluate the performance of ventilation installed in semi-open car parking. The ventilation system is mixed and consists of 16 jet fans and four extractors, as shown in Figure 3.11.

In the following are reported the results for two different operating configurations for the same ventilation system installed in the parking garage:

- first operating configuration, four extractors and the first row of eight jet fans (1 to 8, as shown in Figure 3.11) are in operation;
- second operating configuration, four extractors and 16 jet fans (rows 1 to 8 and 9 to 16, as shown in Figure 3.11) are in operation.

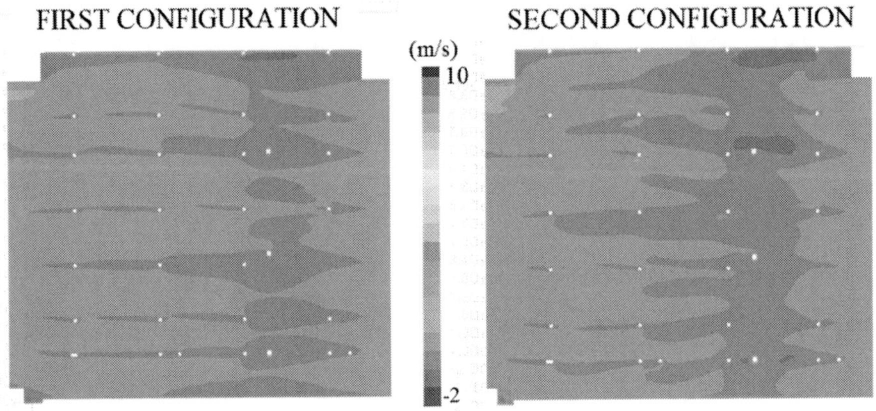

Figure 3.14. Longitudinal velocity field at y=1.5 m for both configurations.

In Figures 3.12 to 3.14 are presented the velocity fields along the plane parallel to the floor that passes through the jet fan symmetrical plane (Figure 3.13), for both configuration, and longitudinal velocity fields at y = 0.5 m and y = 1.5 m (Figures 3.13 and 3.14, respectively) for the first configuration (on the left) and for the second configuration (on the right).

From the Figures, it is possible note that, in both configurations, there are zones of recirculation, in particular in the regions interposed between the rows of fans. This situation is due to the transversal and longitudinal non-optimal distance between, respectively, the two rows of eight jet fans and the two jet fans adjacent in the same row (interference between the flows induced by jet fans).

4. Ventilation Efficiency

4.1. Effectiveness of Local Exhaust Opening

The most effective hood uses the minimum exhaust airflow rate to provide maximum contaminant control. The capture effectiveness should be high, but it would be difficult and costly to develop a hood that is 100% efficient. Makeup air supplied by general ventilation to replace exhausted air can dilute contaminants that are not captured by the hood.

Efficiency of industrial local exhaust ventilation is defined as the ratio of air contaminant quantity captured by the system per unit time to the total contaminant quantity produced by the process per unit time. Today, no direct method exists for the evaluation of the efficiency. Typically, the efficiency evaluation of a industrial local exhaust hood is carried out experimentally, by means of a tracer gas technique (using, for example, sulfur hexafluoride, SF_6).

4.2. Capture Velocity

Capture velocity is the air velocity at the point of contaminant generation upstream of a hood. The minimum hood-induced air velocity necessary to capture and convey the contaminant into the hood is referred to as capture velocity. The contaminant enters the moving airstream at the point of generation and is carried along with the air into the hood. Designers use a capture velocity to select a volumetric flow rate to draw air into the hood. Table 4.1 shows ranges of capture velocities for several industrial operations. These figures are based on successful experience under ideal conditions. Once capture velocity upstream of the hood and hood position relative to the source are known, then the hood flow rate can be determined for the particular hood design. Velocity distributions for specific hoods must be known or determined.

For a given hood configuration and capture velocity, the exhaust volumetric flow rate can be calculated as:

$$Q_0 = V_0 A_0 \tag{4.1}$$

where

Q_0=exhaust volumetric flow rate, m^3/s;

V_0=average air velocity in hood opening that ensures capture velocity at point of contaminant release, m/s;

A_0= hood opening area, m^2.

Low-face velocities require that supplied air be as uniformly distributed as possible to minimize the effects of room air currents. This is one reason replacement air systems must be designed with exhaust systems in mind. Air should enter the hood uniformly; interior baffles are sometimes necessary.

Airflow requirements for maintaining effective capture velocity at a contaminant source also vary with the distance between the source and hood.

Table 4.1. Range of capture velocities

Condition of contaminant	Examples	Capture velocity, [m/s]
Released with essentially no velocity into still air	Evaporation from tanks, degreasing, plating	0.2 to 0.5
Released at low velocity into moderately still air	Container filling, low-speed, conveyor transfers, welding	0.5 to 1.0
Active generation into zone of rapid air motion	Barrel filling, chute loading of conveyors, crushing, cool shakeout	1.0 to 2.5
Released at high velocity into zone of very rapid air motion	Grending, abrasive blasting, tumbling, hot shake out	2.5 to 10

Airflow near the hood can be influenced by drafts from supply air jets or by turbulence of the ambient air caused by jets, upward/downward convective flows, moving people, and drafts from doors and windows. Process equipment may be another source of air movement. For example, high-speed belt material transfer systems, falling granular materials, and escaping compressed air from pneumatic tools produce air currents. These factors can significantly reduce the capturing effectiveness of local exhausts and should be accounted for in exhaust system design.

4.2.1. Hood Capture Velocity (without Flange)

The capture velocity equation depends on the configuration of the hood and duct system. For example, the capture velocity equation (limited for the distance X being less than or equal to 1.5 hood diameters) in the case of hood without a flange, is:

$$Q = V_h(10\,X^2 + A) \qquad (4.2)$$

with Q the necessary exhaust airflow rate, X the distance from hood face to farthest point of contaminant release, V_h the hood capture velocity at distance X and A the area of hood opening.

Example. 1- circular hood opening

Considering a 0.40 m diameter hood, and knowing the recommended capture velocity for a certain pollutant equal to 1.50 m/s, calculate the required volumetric flow rate to guarantee that capture velocity when the distance from the hood is equal to X=0.30 m.

The area of the hood opening is:

$$A = \pi D^2/4 = 3.14 \cdot 0.40^2/4 = 0.13\ m^2$$

Applying the equation (4.2), it is possible to calculate the volumetric flow rate , Q, required to attain the recommended capture velocity:

$$Q = 1.50 \cdot (10 \cdot (0.30)^2 + 0.13) = 1.55\ m^3/s$$

Example. 2- circular hood opening

Considering the precedent example (1), calculate the required flow rate to guarantee the same capture velocity, when the distance from the hood is double (X=0.60 m).

Applying the equation (4.2), it is possible to calculate the new volumetric flow rate , Q', required to attain the recommended capture velocity:

$$Q' = 1.50 \cdot (10 \cdot (0.60)^2 + 0.13) = 5.60\ m^3/s$$

This example shows the distance influence on the flow rate value required to guarantee the capture velocity value imposed: the volumetric flow rate requirements increased (non linearly) when the distance between the hood and the contaminant source increased.

4.2.2. Hood Capture Velocity (Side Baffles or Flanges)

The hood designs for improved performance require the presence of side baffles or flanges that can be used to help block the movement of clean air into the low pressure zone of the hood and, in addition, both help eliminate cross drafts, which can prevent the intended movement of the pollutant-laden gas into the hood.

Another design used to improve capture effectiveness is the push-pull hood, in which a high-velocity clean air stream is blown across the area of pollutant generation into the hood on the opposite side. Push-pull hoods can provide very high capture efficiencies where they are applicable.

The hood capture velocity equation used for a hood with a wide flange is:

$$Q=V_h(0.75)(10X^2+A) \tag{4.3}$$

Example. 3- circular hood opening with a wide flange

Using the same data as presented in Example 1, calculate the required volumetric flow rate to guarantee the capture velocity for a hood with a wide flange.

Applying the equation (4.3), it is possible to calculate the new volumetric flow rate , Q'', required to attain the recommended capture velocity:

$$Q''=1.50\cdot(0.75)(10\cdot(0.30)^2+0.13)= 1.16\ m^3/s$$

It is possible to note that the volumetric flow requirements decreased by approximately 25% when a flanged hood was used.

5. Example of Fire Incident Response in Road Tunnel

5.1. Introduction

An Integrated Traffic and Plant Management System (ITPMS) is used to divert traffic above ground away from the exit slip roads, allowing vehicles downstream of the fire site to be quickly driven away in the event of a fire. The following section discusses the fire incident response adopted for the tunnel.

Step 1. Fire Incident Detection

Prompt detection of a fire in the tunnel is an important factor in preventing a catastrophic fire incident from occurring. The detection criterion is set at 30 to 60 seconds, and Automatic Incident Detector (AID) and linear heat detector are provided.

Step 2. Verification of Fire Incident

When an incident occurs, it is important that the tunnel operator is able to quickly assess the situation and respond to the problem immediately. Real-time information of events occurring in the tunnel is important to the tunnel operator and CCTVs (Closed Circuit

Televisions), and emergency telephones installed in the tunnel provide the means to verify the severity of the tunnel fire. A 60-second timeline is allocated to verify and identify a fire in the tunnel.

Step3. Response to Fire Incident

Upon confirmation of a tunnel fire, depending on the tunnel condition, appropriate response measure includes closure of tunnel by a Variable Message Signal (VMS) to stop additional vehicles from entering the tunnel. Vehicles already in the tunnel will be directed to the nearest exit and vehicles downstream of the fire will be able to drive away. Strobe lights and variable message signs are installed along the tunnel to inform motorists trapped upstream of the fire to evacuate the tunnel. Radio re-broadcast with break-in facilities is also provided to broadcast messages to the motorists. Pre-programmed messages will be broadcast to advise motorists of an incident in the tunnel, and motorists in both the incident and non-incident tunnel would be able to receive the message. The ventilation fans will be turned on to create a longitudinal airflow to prevent backlayering of smoke to the vehicles trapped upstream of the fire. This will create a smoke-free path for the motorists trapped upstream of the fire to evacuate safely through the cross-passage doors. Radio channels used by the emergency services are supported in the tunnel. The tunnel operator will also provide information to the fire service (SCDF) such as the best direction to approach the fire location. Figure 5.1 illustrates the incident-management process when a fire occurs in the tunnel.

The traffic in this tunnel is uni-directional, and a longitudinal ventilation system is adopted for the tunnel ventilation design. The airflow in the tunnel varies depending on the location of the tunnel and the tunnel fan operation mode. In normal operating mode, the tunnel fans are off. Whenever there is traffic congestion in the tunnel leading to a build-up of carbon monoxide (CO), temperature and low visibility (due to emissions from vehicles), the sensors in the tunnel will be triggered, and the tunnel fans will be subsequently turned on.

In the event of a fire, traffic downstream of the fire site will be able to drive away, while traffic upstream of the fire site may be trapped. The tunnel will be ventilated and smoke from the incident vehicles will be released to the atmosphere via exhaust stacks in the ventilation buildings (Figure 5.1). The operation of the tunnel ventilation fans is critical, as it will prevent the spread of smoke and heat to the trapped motorists upstream of the fire site.

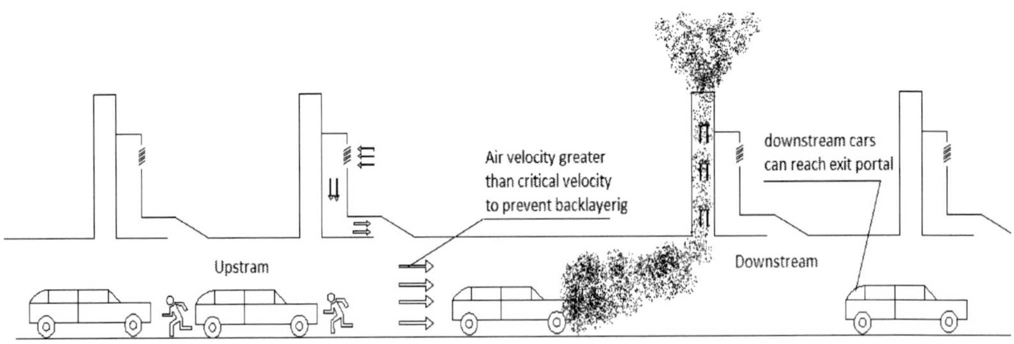

Figure 5.1. Smoke control and evacuation strategy in tunnel.

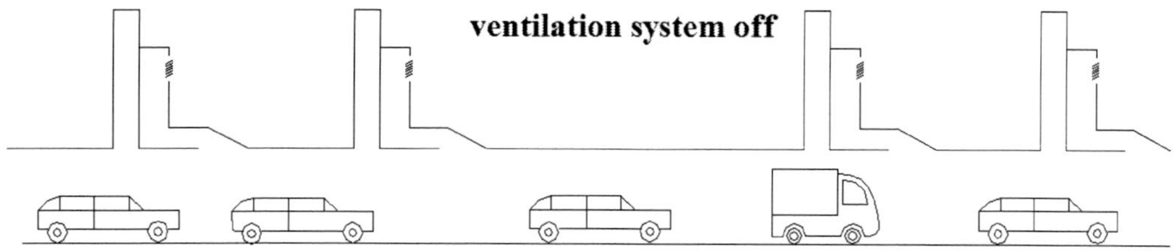

Figure 5.2. a - Ventilation system operating in normal mode. Time after ignition: 0 s, detector and fan status: fan off (no fire detected).

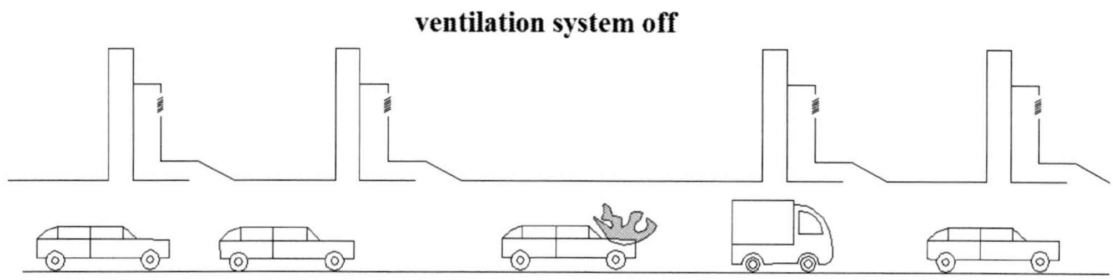

Figure 5.2.b - Ventilation system operating in normal mode. Time after ignition: 60 s, detector and fan status: fire detected, fan off.

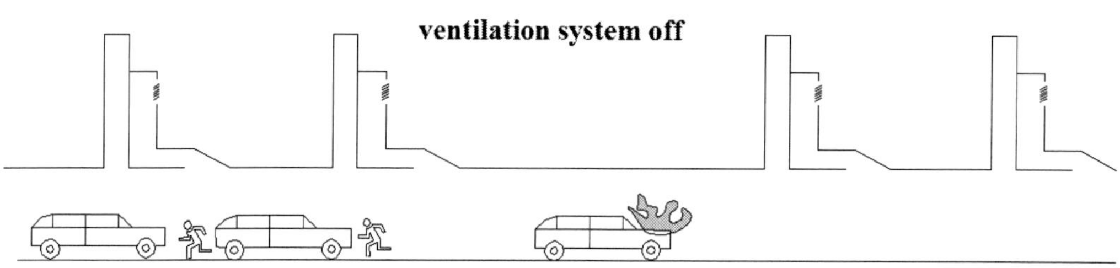

Figure 5.2.c - Ventilation system operating in normal mode. Time after ignition: 120 s, detector and fan status: fire detected, fan off.

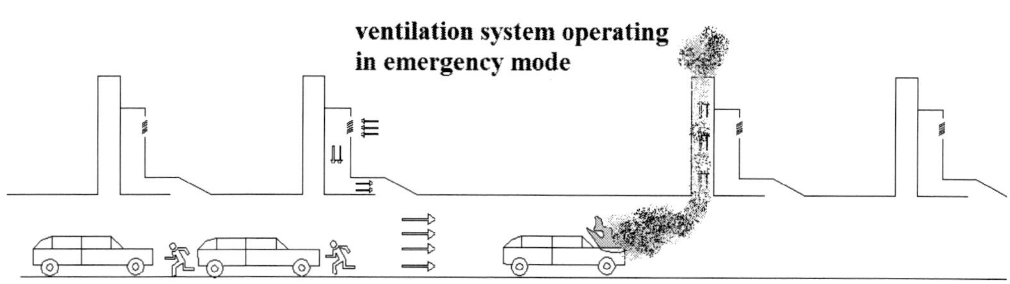

Figure 5.2.d - Ventilation system operating in emergency mode. Time after ignition: 300 s, detector and fan status: fire detected, fan operating in emergency mode.

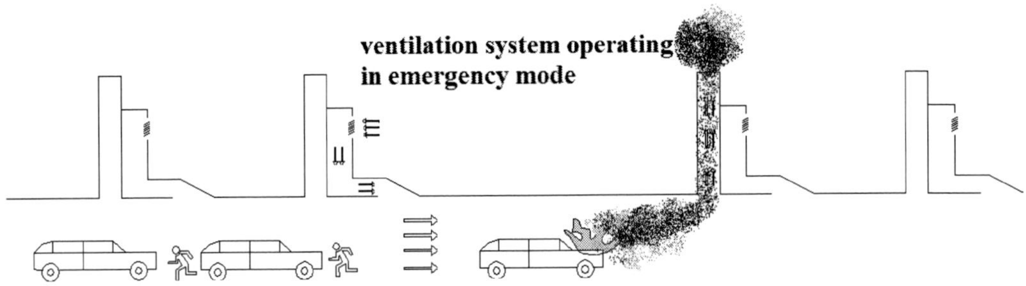

Figure 5.2.e - Ventilation system operating in emergency mode. Time after ignition: 360 s, detector and fan status: fire detected, fan operating in emergency mode at full rotational speed.

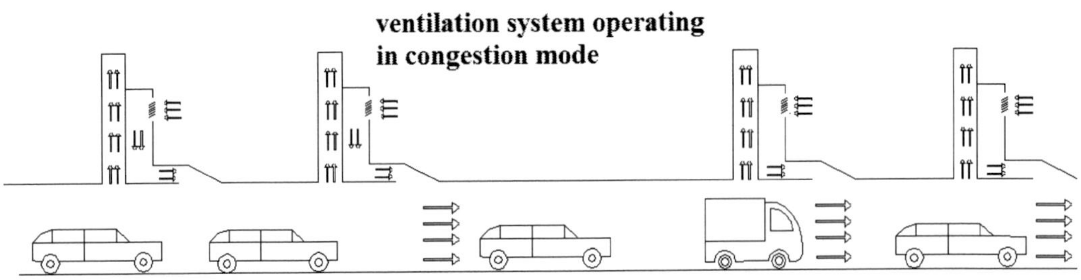

Figure 5.3.a - Ventilation system operating in congestion to emergency mode. Time after ignition: 0 s, detector and fan status: no fire detected, fan operating in congestion mode.

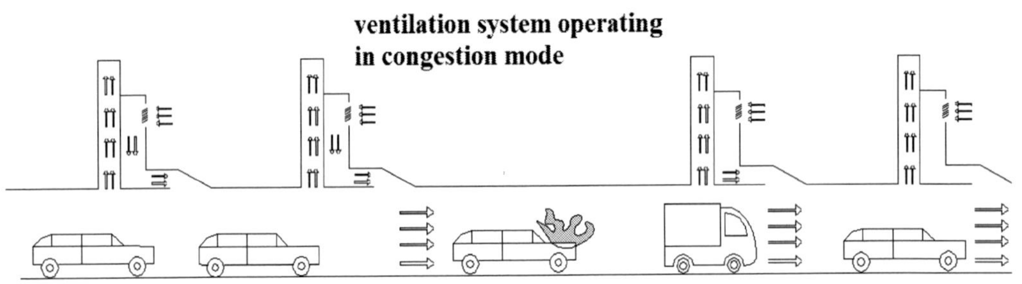

Figure 5.3.b - Ventilation system operating in congestion to emergency mode. Time after ignition: 60 s, detector and fan status: fire detected, fan operating in congestion mode.

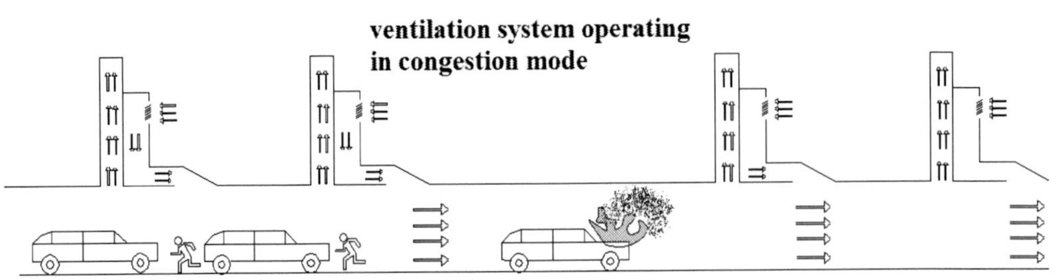

Figure 5.3.c - Ventilation system operating in congestion to emergency mode. Time after ignition: 120 s, detector and fan status: fire detected, fan operating in congestion mode.

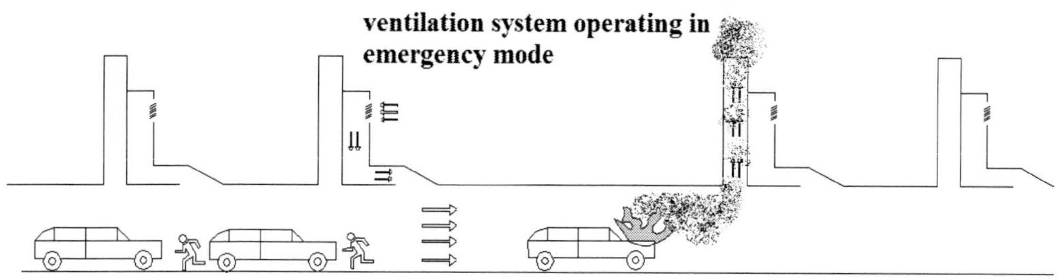

Figure 5.3.d - Ventilation system operating in congestion to emergency mode. Time after ignition: 300 s, detector and fan status: fire detected, fan operating in emergency mode.

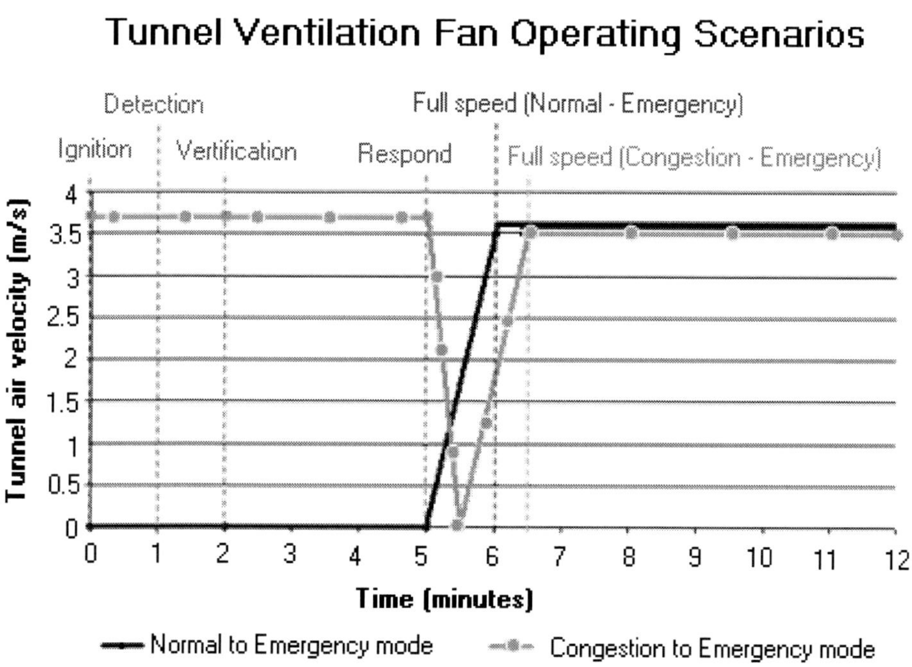

Figure 5.4. Examples of tunnel ventilation fans operating scenarios.

Sufficient airflow is necessary upstream of the fire site to prevent smoke backlayering so that a smoke-free path can be achieved, allowing trapped motorists to evacuate to the non-incident tunnel section.

The design air velocity in the tunnel to achieve critical velocity varies depending on the cross-sectional area and gradient of the tunnel. The tunnel air velocity varies from 1.7 m/s to 5.2 m/s. These airflows are based on the tunnel ventilation design airflow for this tunnel.

Various fire event scenarios considering detection time, operator reaction time and time required for the change of fan operational mode are shown in Figures 5.2 and 5.3.

Typical emergency criteria require the tunnel ventilation fans to operate from standstill to full rotational speed within 60 seconds, or reversible fans completing full rotational reversal within 90 seconds are considered in the following fire events scenarios.

An example of fire scenarios showing the tunnel ventilation fans operating sequence is presented in Figure 5.4.

References

"Guide pratique de ventilation". 1985. Cahiers de notes documentaires, Sècurité et hygiène du travail, 119, pagg. 159-182.

"Industrial Exhaust Systems" - Ashrae Systems Handbook 1973 (Chapter 22, pg. 22.1-22.12).

"Industrial Exhaust Systems" - HVAC ASHRAE Handbook 1987 (Chapter 43, pagg. 43.1-43.10).

"Principles of Local Exhaust Ventilation." 1975. HM Factory Inspectorate, Health and Safety Executive, First Report of the sub-committee on Dust and Fume, Crown Pub., London,.

2007 ASHRAE Applications - Stack Heights and Air Intakes: "Laboratory exhaust stacks should release effluent to the atmosphere without producing undesirable high concentrations at fresh air intakes, operable doors and windows, and locations on or near the building where access is uncontrolled."

ACGIH 2001. Industrial Ventilation, A manual of Recommended Practice 24th Edition.

ACGIH. 1976. Committee on Industrial Ventilation: "Industrial Ventilation - A Manual of Recommended Practice" - 14th Ed., Lansing.

ACGIH. 2001. Industrial ventilation: A manual of recommended practice, 24th ed. Committee on Industrial Ventilation, American Conference of Governmental Industrial Hygienists, Cincinnati, OH.

Alden, J.L. and J.M. Kane. 1982. Design of industrial ventilation systems, 5th ed. Industrial Press, New York.

AMCA. 1995. Fans and systems. Publication 201-95. Air Movement and Control Association International, Arlington Heights, IL.

AMCA. 1995. Field performance measurement of fan systems. Publication 203. Air Movement and Control Association International, Arlington Heights, IL.

ASHRAE HANDBOOK, HVAC Applications, SI Edition, 2003.

Balchin, N.C. (ed.). 1991. Health and safety in welding and allied processes, 4th ed.

Baturin, V.V. 1972. Fundamentals of industrial ventilation, 3rd English ed. Pergamon Press, New York.

Betta V., Cascetta F., Labruna P., Palombo A. 2004. A numerical approach for the air velocity predictions in front of exhaust flanged slot openings, *Building and Environment*, Vol. 39, No.1, January 2004, pp.9-18.

Betta V., Cascetta F., Palombo A. 2007. Push-Pull Ventilation System: a CFD approach for the performance analysis, *International Journal of Ambient Energy*, Volume 28, Issue 3, July 2007, pp.123-134.

Braconnier, R. 1988. Bibliographic review of velocity field in the vicinity of local exhaust hood openings. *American Industrial Hygiene Association Journal,* 49(4): 185-198.

British Occupational Hygiene Society (BOHS). 1987. Controlling airborne contaminants in the workplace. Technical Giade No. 7. Science Review Ltd. and HandH Sci. Consult., Leeds, U.K.

Brooks, P. 2001. Designing industrial exhaust systems. *ASHRAE Journal,* 43(4):1-5.

Caplan, K.J. and G.W. Knutson. 1978. Laboratory fume hoods: Influence of room air supply. *ASHRAE Transactions,* 82(1):522-537.

Cascetta F., Rosano F.M. 2001. Assessment of velocity fields in the vicinity of rectangular exhaust hood openings, *Building and Environment*, Vol. 36/10, November 2001, pp. 1137-1141.

E. Ower, R.C. Pankhurst. 1984. "The measurement of airflow - Pergamon Press Pub., Oxford.

F. Cascetta, L. Bellia. 1996. Velocity fields in proximity of local exhaust hood openings: an intercomparison between current recommended formulas and experimental studies, *Building and Environment*, Vol.31, No.5, September 1996, pp. 451-459.

F. Cascetta. 1996. Experimental evaluation of the velocity fields for local exhaust hoods with circular and rectangular openings, *Building and Environment*, Vol.31, No.5, September 1996, pp. 437-449.

Flynn, M.R. and M.J. Ellenbecker. 1985. The potential flow solution for air-flow into a flanged circular hood. *American Industrial Hygiene Journal,* 46(6):318-322.

Fuller, F.H. and A.W. Etchells. 1979. The rating of laboratory hood performance. *ASHRAE Journal,* 21(10):49-53.

Garrison, R.P. 1977. Nozzle performance and design for high-velocity/low-volume exhaust ventilation. Ph.D. dissertation. University of Michigan, Ann Arbor.

Goodfellow, H. and E. Tahti (eds). 2001. Industrial ventilation design guidebook. Academic Press, New York.

Goodfellow, H.D. 1985. Advanced design of ventilation systems for contaminant control. Chemical Engineering Monograph 231. Elsevier, Amsterdam.

Goodfellow, H.D. 1986. Ventilation '85 (Conference Proceedings). Elsevier, Amsterdam.

H.J. McDermott. 1976: "Handbook of ventilation for contaminant control" - Ann Arbor Science Pub., Ann Arbor.

Hagopian, J.H. and E.K. Bastress. 1976. Recommended industrial ventilation guidelines. Publication No. 76-162.

Heinsohn, R.J., K.C. Hsieh, and C.L. Merkle. 1985. Lateral ventilation systems for open vessels. *ASHRAE Transactions,* 91(1B):361-382.

Hemeon, W.C.L. 1999. Plant and process ventilation, 4th ed. Industrial Press, New York.

Hinds, W. 1982. Aerosol technology: Properties, behavior, and measurement of airborne particles. John Wiley and Sons, New York.

Huebener, DJ. and R.T. Hughes. 1985. Development of push-pull ventilation. *American Industrial Hygiene Association Journal,* 46(5):262-267.

ISO 3966: "Measurement of fluid flow in closed conduits-velocity area method using Pitot static tubes, 1977.

Kofoed, P. and P.V Nielsen. 1991. Thermal plumes in ventilated rooms Vellicai volume flux influenced by enclosing walls. 12th Air Infiltration and Ventilation Centre Conference, Ottawa.

Ljungqvist, B. and C. Waering. 1988. Some observations on "modem" design of fume cupboards. Proceedings of the 2nd International Symposium on Ventilation for Contaminant Control, Ventilation '88. Pergamon. U.K.

M. Shibata, R.H. Howell, T. Hayashi. 1982. "Characteristics and Design method for Push-Pull Hoods" - Part 1, Cooperation Theory on Airflow, *ASHRAE Transaction,* 1982, Paper no. 2697, pagg. 535-556.

M.R. Flynn, M.J. Ellenbecker. 1985. "The Potential Flow Sostition for Airflow into a Flanged Circular Hood" - *Am. Ind. Hyg. Assoc. J.*, Vol. 46, June 1985, pagg. 318- 322.

N.A. Esmen, D.A. Weyel, F.P. McGuigan. 1986. "Aerodynamic Properties of Exhaust Hoods" - *Am. Ind. Hyg. Assoc.*, Vol. 47, pagg. 448-454.

NFPA. 1995. Standard for ovens and furnaces. ANSI/NFPA Standard 86-95. National Fire Protection Association, Quincy, MA.

OSHA, General Industry Safety and Health Regulations. 1975- US Code or Federal Regulations, Title 29, Chapter XVII, Part 1910.1006.

Posokhin, V.N. and A.M. Zhivov. 1997. Principles of local exhaust design. Proceedings of the 5th International Symposium on Ventilation for Contaminant Control, Vol. 1. Canadian Environment Industry Association (CEIA), Ottawa.

Qiang, Y.L. 1984. The effectiveness of hoods in wìndy conditions. Kungl Tekniska Hoggskolan. Stockholm, Sweden.

R. Braconnier, G. Aubertin. 1983. "Captage et dilution des pollants," Travail and Sécurité, 1-2, pagg. 26-42.

R.H. Kirchhoff. 1985. "Potential Flows-computer graphic solutions" - Pub. Marcel Dekker Inc., New York,.

Safemazandarani, P. and H.D. Goodfellow. 1989. Analysis of remote receptor hoods under the influence of cross-drafts. *ASHRAE Transactions,* 95(1):465-471.

Sciola, V. 1993. The practical application of reduced flow push-pull plating tank exhaust systems. 3rd International Symposium on Ventilation for Contaminant Control, Ventilation '91, Cincinnati, OH Chambers, D.T. 1993.

Sepsy, C.F. and D.B. Pies. 1973. An experimental study of the pressure losses in converging flow fittings used in exhaust systems. Document PB 221 130. Prepared by Ohio State University for National Institute for Occupational Health.

Shibata, M., R.H. Howell, and T. Hayashi 1982. Characteristics and design method for push-pull hoods: Part I—Cooperation theory on airflow; Part 2—Streamline analysis of push-pull flows. *ASHRAE Transactions,* 88(l):535-570.

SMACNA. 1999. Round industrial duct construction standards, 2nd ed. Sheet Metal and Air Conditioning Contractors' National Association, Chantilly, VA.

V. Betta, F. Cascetta, L. Bellia. 1996. New Empirical Equations for the Calculation of Centerline Air Velocity in Front of Exhaust Openings, *ASHRAE Transactions,* 1996, Vol.102, Part 1, pp.230-238.

V. Betta, F. Cascetta, L. Bellia.1998. Description of velocity fields in front of exhaust axisymmetric inlets by means of new empirical equations, *Building and Environment*, Vol.33, No.4, 1998, pp.189-195.

Vittorio Betta, Furio Cascetta, Marilena Musto, Giuseppe Rotondo. Numerical study of the optimization of the pitch angle of an alternative jet fan in a longitudinal tunnel ventilation system. *Tunnelling and Underground Space Technology*, Vol. 24 pp. 164-172, March 2009.

Zarouri, M.D., R.J. Heinsohn, and C.L. Merkle. 1983. Numerical computation of trajectories and concentrations of particles in a grinding booth. *ASHRAE Transactions,* 89(2A):119-135.

In: Ventilation: Types, Standards and Problems
Editors: V. A. Romano and A. S. Duval, pp. 91-109
ISBN 978-1-61324-424-1

Chapter 3

PULMONARY VENTILATORS AS MONITORING INSTRUMENTS: MEASUREMENTS RELIABILITY AND COMMON ISSUES

Sergio Silvestri*

Measurements and Biomedical Instrumentation, Faculty of Biomedical Engineering "Campus Bio-Medico", Rome, Italy

ABSTRACT

The high prevalence of respiratory diseases has led to an increasing demand of assistance services in recent years. This increase also introduced a gradual rise in the admission of patients, what it is particularly important in the case of the respiratory diseases, whose prevalence is high and expected to grow in a near future. Acute lung injury, acute respiratory failure and its most severe form, the acute respiratory distress syndrome, are relatively common in the intensive care unit setting and have a high morbidity and mortality. Chronic obstructive pulmonary disease, for example, affects about 210 million people globally and causes some 3 million deaths annually. These diseases are a major drain on healthcare budgets with 50% of costs accounted for by hospital admissions, much of which could be avoided through development of more responsive models of care allowing earlier recognition and a better treatment.

Almost all most critical patients admitted in a Critical Care Unit necessitate ventilatory support. Despite a research shift to non-invasive respiratory support, mechanical ventilation remains an essential tool in the care of the critically ill. The availability of a variety of respiratory modes and somewhat confusing terminology presents a daunting challenge for the physician. But, above all, many of the available treatment modes have not been adequately evaluated and their efficacy appears not completely assessed, by only judging the extent of recent literature reporting experimental evaluations *in vivo* and *in vitro*. In addition, there is paucity of information on the relative merits of those ventilatory strategies that have been studied. From an analysis of recent literature, mechanical, pneumatic, fluid-dynamic and thermo-hygrometric phenomena taking place in the system connecting the patient to the

* Via Alvaro del Portillo, 21 – 00128, Rome, Italy. Ph. +39-06-225419604, Fax.+39-06-225419609. E-mail: s.silvestri@unicampus.it.

pulmonary ventilator (breathing circuit) seem to influence, and in a not always repeatable way, the pneumatic parameter values actually applied to the patient, the thermo-hygrometric state of the gas mixture, the response time to patient's inspiration effort and the assessment of patient respiratory parameters.

Just to give an example, in the last decade, measurement of gas volume (air and oxygen) actually delivered to patient's lungs during mechanical ventilation, especially in case of neonatal ventilation, has received the attention of several scientific authors. The main outcome of this research is that the value of the delivered volume displayed by the ventilator monitoring unit significantly differs from the actual volume that reaches patient's airways. These observations assume increasing importance because of the low mortality associated with ventilation with reduced tidal volumes.

It has to be considered that mechanical ventilation is always an interactive process between the machine (pulmonary ventilator, humidifier, breathing circuit) and the patient (characteristics of the respiratory system, kind of patient and pathology, *etc.*), also if the patient is completely passive. Thus, it appears interesting to investigate the above described measurement systems for monitoring and control of mechanical quantities, as it is common practice with other instruments utilized in other fields of science. Moreover, several parameters of patient's pulmonary mechanics are deduced from these measurements. Thus, clinicians figure out quantitative evaluation of the mechanical status of the respiratory system, i.e. respiratory system compliance, airway resistance, work of breathing, from numerical quantities provided by the ventilator monitoring unit and from the observation of the shape of respiratory loops. However, there is no assurance that the air flow and pressure actually delivered to the patient as a function of time correspond to the respiratory pattern set by the physicians.

Probably, in order to better understand the reliability and clinical significance of monitoring results, it is necessary to address the study of pulmonary ventilation by considering the system ventilator-circuit-patient as a whole.

This chapter, which is intended to represent a food for though for the clinician and researcher dealing with pulmonary ventilation, has the aim to briefly address some causes of possible misleading information which can be obtained from these complex though fascinating machines for vital support.

INTRODUCTION

The interest for biological injury induced and associated to ventilatory therapy has been gradually growing in the last years, until becoming one of the main topics in current scientific and medical literature in the fields of resuscitation and anaesthesia [1, 2, 3]. Although experimental evidences correlating pathologies of the pulmonary parenchyma and effective mechanical ventilation strategies are still poor, the potential harmful effects of ventilatory support appear relevant in patients with acute respiratory insufficiency (Acute Lung Injury - ALI, Acute Respiratory Distress Syndrome - ARDS) [4, 5]. In the last years, scientific research in this field obtained remarkable results, due to the large quantity of studies having identified the risk factors and potentially dangerous effects of several mechanical ventilation modalities. Therefore, a vast part of literature concentrates on developing therapeutic strategies in order to avoid the pulmonary damage induced by ventilation.

Webb and Tierney [6] appear the first, in 1974 through a study on animals, to show without ambiguity that mechanical ventilation is able to produce pulmonary edema. Since then, the scientific community chose to classify the pulmonary damage in two macro-pathologies, called *barotrauma* and *volutrauma*, being the first associated to macroscopic phenomena, like the air presence in extra-alveolar space, and the second one associated to microscopic alterations, like the increase of vascular permeability [7]. The classification of induced pulmonary damage in the two above-mentioned categories survived for many years, until more recent research showed that the two terms would seem referring to coexisting, and not always easily distinguishable, pathological situations [8, 9]. Currently, in order to refer to potential harmful effects due to ventilatory therapy, the generic term Ventilator Induced Lung Injury (VILI) is preferred [10].

In order to avoid the induced pulmonary damage, several ventilation strategies have been developed: in some cases through the combination of traditional and partial liquid ventilation, or more simply, by reducing the tidal volume with the consequent reduction of positive pressure applied to lung parenchyma. These ventilation techniques, e.g. permissive hypercapnia and low-volume ventilation, have lead to a significant reduction of iatrogenic effects and pathologies associated with ventilatory treatment, in patients affected from acute respiratory insufficiency, and today they are known as *protective ventilatory strategies*. As a direct consequence, the renewed attention of medical scientific community has been directed to the monitoring of the mechanical parameters of the respiratory system, whether they be static or dynamic, being these among the preferential indices for the assessment of an appropriated ventilatory therapy.

However, the availability of a variety of respiratory modalities and somewhat confusing terminology presents a daunting challenge for the physician. But, above all, many of the available treatment modes have not been adequately evaluated and their efficacy appears not completely assessed, by only judging the extent of recent literature reporting experimental evaluations both *in vivo* and *in vitro*. In addition, there is paucity of information on the relative merits of those ventilatory strategies that have been studied.

The study of innovative modalities in this field, in order to be effective, is based on the reliability of the measured parameters characterizing the mechanics of ventilation. The measurements of these parameters, as already mentioned, are usually performed by a monitoring unit through some transducers. Sensors are usually incorporated into the ventilator device, performing measurements, generally, where the gases are not humidified and at ambient temperature and, therefore, very often, upstream the patient circuit. Experimental evidence shows that the values of respiratory parameters performed by the ventilator are different if compared with same measurements measured by other monitoring devices. A misunderstanding in the evaluation of parameters characterizing the mechanics of respiration may play a significant role on the efficacy evaluation of the ventilatory strategy.

From an analysis of recent literature, mechanical, pneumatic, fluid-dynamic and thermo-hygrometric phenomena taking place in the system connecting the patient to the pulmonary ventilator (breathing circuit) seem to influence greatly, and in a not always repeatable way, the pneumatic values applied to the patient, the thermo-hygrometric state of the gas mixture, the response time to the patient's inspiration effort *et cetera*.

In order to analyze the several discrepancies and the rather reduced repeatability reported in literature, we suppose that the effects of ventilatory system pneumatics, combined with ventilation modes, play a primary role in the assessment of actual pneumatic parameters applied to the patient.

The Patient – Pulmonary Ventilator System

A pulmonary ventilator is a machine able to insufflate a known gas quantity in the lungs and, subsequently, to allow the expiration, with known frequency and an appropriate pressure waveform. The term "ventilator" has a mechanical origin, since it is an operating machine which delivers energy to the gas with low prevalence; according to the classic distinction where a merely quantitative difference distinguishes a ventilator from a compressor, which is defined as an operating machine delivering energy to the gas with high prevalence. Therefore, the pulmonary ventilator can be schematized as a generator of mechanical energy supplying energy to a load (the patient).

This simple observation is the basis of the electrical analogy with which, in the scientific literature, the patient-pulmonary ventilator system is frequently modeled. Such a schematization is useful in order to understand the modalities of controlled ventilation strategies, being they pressure or volume controlled strategies, since they are just different modalities for energy transfer from a generating system to a load. Moreover, the analogy allows to understand the pressure distribution inside the entire system utilized in mechanical ventilation.

The analogy consists in representing the pulmonary ventilator as a voltage generator, e.g. a battery, and the patient as an electrical load, composed by an electrical resistance and capacitance, in order to schematize the resistance to gas flow rate and the elasticity of the respiratory apparatus. With reference to figure 1, it can be observed that the electrical analogy represents the gas flow rate - $\dot{V}(t)$ - with the electrical current I(t), and the alveolar pressure $P_{alv}(t)$ with the voltage E(t). This last parameter is measured downstream the airways resistance (R_{RS}) and allows to separate airways fluid-dynamic resistance (R_{RS}) from the parameter representing the elastic contribution, i.e. the pulmonary compliance (C_{RS}).

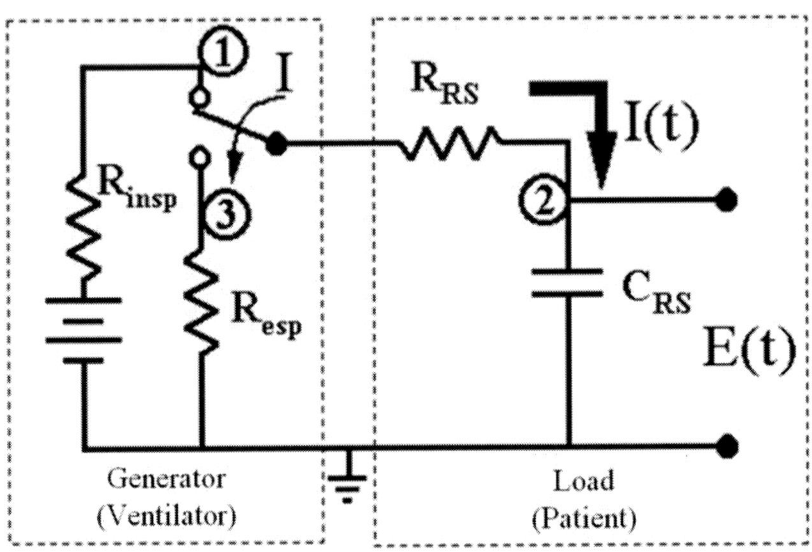

Figure 1. Schematic representation of the ideal electrical analogy patient-pulmonary ventilator.

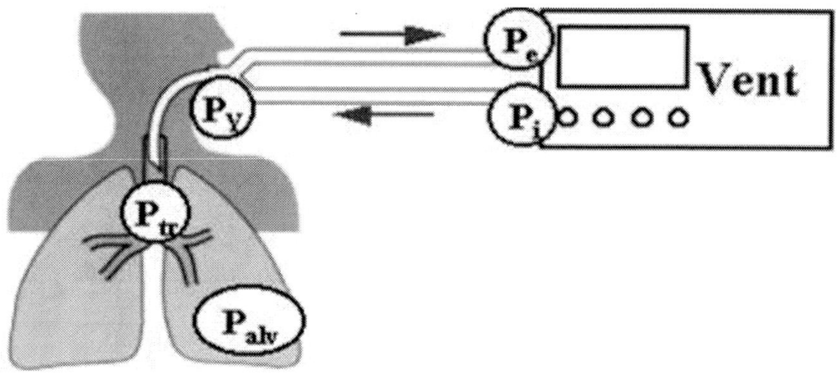

Figure 2. Schematic representation of pressure distribution along the patient-ventilator system during transient conditions. P_i, pressure upstream the circuit inspiratory limb; P_e, pressure downstream the circuit expiratory limb; P_Y, pressure at Y-piece; P_{tr}, tracheal pressure; P_{alv}, alveolar pressure.

The transition from inspiration to expiration is timed by the switch (I) that, during inspiration, connects the generator with the patient through the breathing circuit. The breathing circuit, in this extremely simplified analogy, is represented by the fluid dynamic resistances (R_{insp} and R_{esp}) of the inspiratory and expiratory limbs respectively. Of course, in the case of controlled ventilation, the patient is completely passive, that is, there are no other generators able to supply energy. As one might guess, this basic model could be further enriched in order to simulate more complex situations or specific diseases.

The Estimation of Airway Pressure

The above described physical model is an extremely simplified and ideal one, in which many aspects have been intentionally neglected however, also such a simple model allows to draw some observations regarding the monitoring of such an important parameter like airway pressure. In fact, in dynamic conditions, i.e. during the transient time, corresponding to the inspiratory and expiratory phases, the voltage E(t), measured at point 2 in figure 1, is not the same voltage which would be measured at point 1 or at point 3. In particular, just in order to allow the current flow, the voltage in point 1 will be necessarily higher than the voltage in point 2 and, during expiration, the voltage in point 2 will be higher than the voltage in point 3, until the equilibrium will be reached, i.e. at end inspiration or end expiration. Thanks to the analogy, the corresponding pressures will have analogous spatial distribution.

This simple observation allows to understand how to deal with a first misleading information about the monitoring of pressure. Pulmonary ventilators are normally equipped with sensors to measure gas pressure: pressure values, waveforms and trends are continuously displayed on the ventilator monitor. However, in transient conditions, i.e. during the vast majority of a respiratory act time, the value displayed depends on the particular location of the ventilator-patient system where the measurement is performed. Most commercially available ventilators, in fact, measure pressure at the ventilator level, i.e. upstream or downstream the ventilation circuit, mainly because the gas mixture, being mostly dry, allows a more reliable measurement and the sensors require less maintenance. Usually, as far as it concerns the measurement of pressure, two transducers are present, generally positioned as reported in figure 2, inside the ventilator, upstream the inspiratory limb (P_i) and downstream the expiratory limb (P_e) of the breathing circuit respectively. The waveform of Y-piece pressure (P_Y), whose value is visualized on ventilator display, which should be representative of the tracheal pressure (P_{tr}), is normally a calculated rather than a measured value. And, in particular, it is calculated, as it will be shown further on, from the measured values of flow rate and pressure obtained from the sensors P_i and P_e.

The algorithm, implemented in the software of the greater part of ventilators currently in commerce, presumes that, during artificial ventilation, the temporal variation of pressure at Y-piece (P_Y) follows the variations of the pressure P_i, minus a pressure loss due to the circuit inspiratory limb fluid-dynamic resistance (R_{insp} in Figure 1). Therefore, the value of $P_Y(t)$ can be obtained by the formula:

$$P_Y(t) = P_i(t) - R_{insp} \cdot \dot{V}(t) \qquad (1)$$

being $\dot{V}(t)$ the volumetric flow-rate, as a function of time, measured by a flow transducer usually positioned upstream or downstream the patient circuit. For the application of equation (1), the knowledge of R_{insp} value is necessary, i.e. the fluid dynamic resistance of the inspiratory limb of the patient circuit. Therefore, its value is obtained during the ventilator calibration procedure as a value proportional, typically greater than 60 percent, of the total fluid dynamic resistance of the breathing circuit (R_{TOT}). In fact, during the calibration procedure, previous to the ventilator utilization, the operator blocks the common channel of the Y-piece while the machine delivers one or more steady and known flow rate values, usually equal to a few liters per minute, so as to obtain the value R_{TOT} from the application of the formula:

$$R_{TOT} = \frac{P_i - P_e}{\dot{V}_{\cos t}} \qquad (2)$$

As already mentioned, once the value of the total fluid dynamic resistance of the patient circuit (R_{TOT}) is known, the value R_{insp} is calculated as a fraction of R_{TOT}, generally equal to 65-75%. In the above described procedure, whereby the ventilator calculates the Y-piece pressure from the values measured upstream the patient circuit, there are two implicit assumptions: (a) the pressure drop along the circuit varies linearly with the flow (Equation (1)), (b) the value of R_{insp} is constant and independent form the flow rate. These assumptions, generally speaking, are not verified in a tubing system like the one utilized in circuits for artificial ventilation, but the manufacturer guarantees, according to tests conducted at factory level, that the error in the calculation of the pressure is generally lower than 1 or 2 cmH_2O. Please note that the validity of the above calculation procedure is subject to the fact that the ventilator is equipped with breathing circuits with resistance values that are within the range required by the manufacturer of the machine. If the circuit utilized does not comply with these limits, the calculation does not provide a reliable value for the pressure at the Y-piece with the accuracy stated by the manufacturer, and provides numerical values with errors also significantly higher than those above reported.

From the above reported consideration, it is clear that pulmonary ventilators with pressure sensors placed inside the machine are able, albeit with some approximation, to provide the pressure at the Y-piece but, in any case, never the value of pressure in the trachea (P_{tr} in figure 2), this depending on the pressure drop of the intubation device, or even less than the alveolar pressure (P_{alv} in figure 2), although this last is the parameter of most interest being directly associated with the lung injury [11].

Although a single value representative for alveolar pressure is not theoretically possible due to the anisotropy and the complex branched structure of the lung parenchyma, which would not make sense to indicate a single value for the alveolar pressure but rather a map of pressure distribution, tracheal pressure is certainly a more efficacious parameter than the pressure at the Y in order to monitor the dynamic characteristics of the respiratory system [12]. Moreover, it should also be reported that research has shown substantial differences between the pressure volume loop (PV loop) obtained from pressure measured at the Y-piece with respect to the same loop obtained from pressure measured directly in trachea [13]. These differences become even more relevant in the case of partial obstruction of the endotracheal tube. With secretions increase, the pressure P_Y increases significantly even though there is a small corresponding increase in tracheal pressure. This means, obviously, even a considerable error in the calculation of *work of breathing* (WOB), being this parameter proportional to the area enclosed within the PV loop.

RESPIRATORY PARAMETERS EVALUATION PERFORMED BY THE VENTILATOR

In the simple electrical analogy cited in the beginning of the previous paragraph an important detail has been neglected for simplicity: the presence of the breathing circuit. The connection load-generator, as well as all electrical connections, is not a direct connection but the two elements are usually connected through another system, in electrical engineering the connection system is called *transmission line* and, in the case of mechanical ventilation, is represented by the patient circuit. Any energy transfer, whether it be mechanical or electromagnetic one, also shows two characteristic phenomena: (a) it never happens at "zero cost", namely the energy is never fully transferred from the generator to the load, i.e. a part of it is dissipated; (b) any kind of generator-load system has the essential mean of connecting the two elements, in our case the breathing circuit, therefore it can give rise to phenomena of energy reflection, storage and loss.

The main function of the connection is to allow the maximum energy, or power, transfer from the generator to the load, by minimizing loss and reflection, thus realizing what, in technical language, is called *impedance matching*. A typical example of impedance matching is constituted by the gel utilized in diagnostic ultrasound, in which case the gel eliminates the presence of air between the skin and the transducer, reducing the high acoustic impedance due to air presence, and thus assisting the energy transfer from, in this case, the ultrasonic generator (transducer) and the load (internal organ to be inspected). When considering the monitoring of respiratory parameters during artificial ventilation, the presence of the patient circuit cannot be neglected, this would lead to significant errors in measurement results, since the sensors, placed inside the system for conveying gas, measure physical quantities whose value is determined by the entire ventilation system consisting of breathing circuit, the humidifier, condensate drain traps, filters, valves, *et cetera*. In addition, as far as it concerns this subject, a very limited number of experimental studies is present in current scientific literature.

In order to have an idea of the magnitude of this phenomenon just consider, for example, that the tidal volume delivered to the patient (V_{TP}) is not constant with the increase of airway pressure, which can result from a reduction compliance of the respiratory system (C_{SR}), an increase of resistance or both events. The breathing circuit, in fact, is the place where the measurement of respiratory parameters is performed and it is constituted of elements showing an elastic behavior, mainly because of air compressibility. Therefore, the breathing circuit behaves like a spring, capable of storing energy, and performs as an additional compliance, setting up the so-called added *dead space*.

Measurement of the gas volume actually delivered to patient's lungs during mechanical ventilation has recently received the attention of several scientific authors [14, 15, 16, 17]. The main outcome of this research is that the value of the tidal volume displayed by the ventilator monitoring unit significantly differs from the actual tidal volume that reaches the airways of the patient, the value of which could be measured by a pneumotachometer placed at the airways opening. In fact, most commercially available ventilators measure the tidal volume at the ventilator level (V_{TV}), i.e. upstream or downstream of the ventilation circuit, mainly because the gas mixture being mostly dry the measurement is more reliable and the flow sensors require less maintenance, although this way of measuring causes all the measurements to be affected by the pneumatic behaviour of the breathing circuit which connects the patient to the machine [18, 19, 20]. In order to estimate the value of the actual tidal volume (V_{TP}) from measurements performed at the ventilator level, it is usual to compensate for the effects of the breathing circuit by means of the following equation:

$$V_{TP} = V_{TV} - C_i \cdot (PIP - PEEP)\,. \qquad (3)$$

where C_i is the compliance of the breathing circuit, and PIP and PEEP are the Peak Inspiratory Pressure and the Positive End Expiratory Pressure, respectively. Algorithms for the calculation of the V_{TP} based on equation (3) are also implemented in the software of pulmonary ventilators.

However, in many cases encountered in the literature this method appears not to be completely reliable, or its efficacy seems limited to particular cases [21]. An *in vitro* study [22] showed that tidal volume measurements performed by three ventilators and corrected by the application of equation (3) differed markedly compared with the ones performed at the Y piece by two freestanding pulmonary monitors, leading to 95% overestimation and 25% underestimation errors, these being the greatest errors in measurements taken at the smallest tidal volumes and with less compliant test lungs. More recent *in vivo* [23] and *in vitro* [24] research has shown that the overestimation errors obtained by the application of equation (1) are inversely related to the respiratory compliance and the tidal volume, i.e. the lower the tidal volume the higher the error.

These observations assume increasing importance because of the low mortality associated with ventilation with reduced tidal volumes. Thus, the measurement accuracy and reliability of the actually delivered tidal volume is of great interest for ongoing scientific research on this topic and subsequent clinical applications [25].

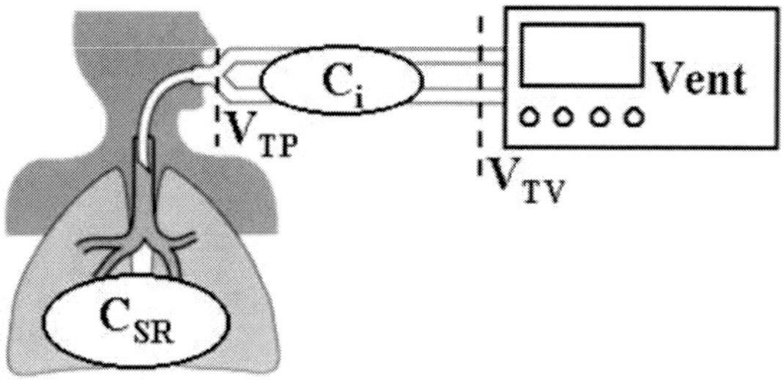

Figure 3. Schematic representation of the patient-circuit-ventilator system. C_i and C_{SR} are the compliance of the breathing circuit and the patient, respectively. V_{TV} and V_{TP} are the tidal volume delivered by the ventilator and the tidal volume reaching patient airways, respectively.

In order to have an idea of the breathing circuit volume we must consider that the volume of a tubing system, with 2.0 cm inner diameter and length of about 1.5 m, encloses, in total, a volume of air equal to approximately 2 liters, excluding the humidifier and water traps. The elastic effect of the patient circuit is mainly determined by the gas compressibility, however, some patient circuits show also small variations of their geometric features during ventilation, which can give origin, sometimes, to vibrational phenomena. These elements influence the value of the gaseous mixture volume that effectively reaches patient airways. The gas in the circuit, in fact, is compressed during inspiration and expands during expiration thus showing, during inspiration, a gas density increase which does not participate to the respiratory exchanges. An increase of pressure, as a consequence, determines a greater percentage of V_T compressed and "lost" in the circuit, with the consequence that a smaller volume reaching the alveolar surface.

Wilson and co-authors [26] were the first to propose eq. (3) as a mean to calculate the tidal volume that actually reaches patient airways from the knowledge of other parameters. Equation (3) arose from the consideration that the volume that actually reaches the patient's airway is equal to the volume delivered by the ventilator minus the volume compressed in the circuit, which is directly proportional to the compliance of the circuit and the maximum pressure increase. However, already Wilson *et al.* and, later Cannon *et al.*, found little correlation between the results obtained from the application of eq. (3) and the volumes experimentally measured using a pneumotachometer placed immediately upstream of the trachea. Therefore, they concluded that the only effective methodology to know the V_{TP} requires the introduction of an additional device. Probably, in order to understand the reasons behind these results, it is necessary to address the study of the volume delivered by considering, once again, the system ventilator-circuit-patient as a whole.

With reference to figure 3 it can be observed, in fact, that the “loads” of the entire system consist of both the patient circuit, with all its accessories whose compliance is shown in the figure with C_i, and the patient respiratory system, which is in series to the circuit and is characterized by compliance C_{SR}. The tidal volume delivered by the ventilator (V_{TV}) will be distributed in proportion to the various compliances, as is the case by applying a weight to two springs in series and suspended vertically at one end: their deformations will be, of course, directly proportional to the respective elastic constants. From the observation of Figure 3, patient compliance C_{SR}, must play a role in the distribution of the volume. Therefore, the volume reaching patient airways (V_{TP}) will be, because of the breathing circuit compliance (C_i) added in series, a fraction of the volume delivered by the ventilator, whose numerical value can be expressed from the formula:

$$V_{TP} = \frac{C_{SR}}{C_{Tot}} V_{TV} \tag{4}$$

where $C_{Tot} = C_i + C_{SR}$, being the two elastic elements in series. Therefore, with reference to figure 3:

$$V_{TP} = V_{TV} \frac{1}{1 + \dfrac{C_i}{C_{SR}}} \tag{5}$$

Just to give an example, if a patient with C_{SR}=15 ml/cmH_2O is ventilated through a breathing circuit with compliance C_i=1 ml/cmH_2O, setting on the ventilator a tidal volume V_{TV}=500 ml a V_{TP}=468 ml will reach patient airways, i.e. about 94%. The lower the difference between the compliance of the breathing circuit and the compliance of the patient, the lower will be volume actually delivered. For C_i=C_{SR} just one half the volume reaches patient airways. The calculation is obviously approximate because we hypothesized constant compliances, whereas this physical parameter is pressure-dependent and non-linear however, it can be useful in order to get an idea of the orders of magnitude associated with this phenomenon. Equation (5) shows that the V_{TP} value depends on the ratio of the compliances following an approximately hyperbolic function, meaning that if C_{SR} is very small compared to C_i, i.e. for small infants, a little variation in this ratio produces a high variation in the ratio of the volumes. This observation might explain the high variability of the results reported in the literature when applying the correction expressed by equation (5) to infants.

However, being compliance mainly due to gas compressibility, the rapidity at which insufflation is performed, i.e. the flow waveform, appears to have a not negligible influence on a reliable applicability of eq. (5) or of other algorithms to indirectly estimate actual tidal volume. As outlined in a research on mechanical simulation of preterm infant lungs [27], the amount of gas ventilated within each respiratory act depends, in a not negligible way, on the rapidity of the delivery. In principle, there are only two situations where gas compressibility can be theoretically assessed and, therefore, tidal volume accurately estimated: either no heat is dissipated through the control volume wall (adiabatic condition) or all the produced heat is dissipated (isothermal condition). As it appears virtually impossible to build a ventilation system that fulfils one of these two extreme conditions, in most cases pulmonary ventilation will behave somewhere between the isothermal and the adiabatic condition. As a consequence, there will be an uncertain thermal behavior of the gas, which may lead to underestimation or overestimation of the actual tidal volume depending on how fast the gas is delivered.

Volumetric Flow Rate, Mass Flow Rate or Volume Variation?

One of the difficulties encountered when approaching the study of respiratory mechanics is the clear distinction between these parameters: volumetric flow rate, mass flow rate and volume variation. This issue was also analytically dealt with by Ligas and Primiano [28]. It seems convenient to point out a simple but not obvious notion that probably appears of little importance in the clinical practice, but has the power to compromise the understanding of some basic concepts in respiratory mechanics. In fluid dynamics, the volumetric flow rate (Q) is defined as the volume of fluid which passes through a given surface per unit time. This parameter is measured in unit of volume per time unit, i.e. liters/s or liters/min or cubic meters per second. On the other hand, the mass flow rate ($\dot{m}$) is defined as the mass of a fluid (or solid) which passes through a given surface per unit time, it is measured in kilograms per second [kg/s] or kilograms per minute [kg/min]. For incompressible fluids, the mass flow rate can be obtained by multiplying the volumetric flow rate by the density of the fluid:

$$\dot{m} = \rho Q \qquad (6)$$

being ρ, the density of the fluid, measured in mass per volume unit, e.g. kg/liters or equivalent. Fluid dynamics assert that whichever fluid, liquid or gas, flowing inside a tube, in absence of fluid drains or sources, preserves its mass flow rate (conservation law). In case of liquids, being these incompressible fluids (ρ=constant), the conservation law extends to the conservation of the volumetric flow rate. However, this is not generally true in case of motion of compressible fluids, such as air and gases, for what the conservation law applies only to mass but not necessarily to volume. In pulmonary ventilation it is, therefore, misleading to think that volumetric flow rate supplied by the ventilator, and measured upstream the patient circuit, is equivalent to the variation of pulmonary volume per unit time ($\dot{V}$), although the gas flow rate is usually reported with this symbol. It is therefore also misleading to integrate this parameter, in order to obtain a measurement of pulmonary volume variation. Such a procedure would be obviously correct if the fluid was a liquid, but in the case of gases the compressibility must be taken into account. If we think, as an example, to the case of extremely rigid lungs it is observed that, by insufflating a known gas volumetric flow rate, the pulmonary volume would remain constant, whereas the gas pressure inside would increase. Naturally, in such a case, the kinetic energy of the fluid would not produce a mechanical work, i.e. a pressure variation multiplied by a volume variation, but it would be 'stored' in the gas as pressure and/or temperature increase. This is due to the mass conservation law, in fact, if the container volume cannot increase then the gas density will have necessarily to increase, with a consequent increase of pressure and temperature.

This observation can be theoretically justified by considering the geometric volume of the lungs equal to the volume of the gas mass in certain conditions of temperature and pressure. Therefore, the perfect gas law can be applied:

$$P \cdot V = n \cdot R \cdot T \qquad (7)$$

where P is the gas pressure inside the lungs (P_{alv}), V is the lung volume, n the number of gas moles in the lung, R the universal constant for gases (8,314 $J \cdot mol^{-1} \cdot K^{-1}$) and T the gas temperature. In case of air, and remembering that air is a gas mixture whose molar mass is approximately equal to: MM_{air}=28.96 g/mol, equation (7) can be written as:

$$P \cdot V = m \cdot R_{air} \cdot T \qquad (8)$$

where $R_{air} = R/MM_{air} = 0{,}287 \ J \cdot g^{-1} \cdot K^{-1}$, m is the air mass in the lungs in grams.

During pulmonary ventilation, the lungs are not a closed system, they exchange mass with the outside, therefore, by assuming as simplifying hypotheses that the entering air has the same temperature of air in the respiratory system, equation (8) can be differentiated over time obtaining:

$$P \cdot \dot{V} + \dot{P} \cdot V = \dot{m} \cdot K \qquad (9)$$

having indicated with K the $R_{air} \cdot T$ product, being this one constant. The mass conservation law applies also to compressible fluids, therefore, from equations (6) and (9):

$$P \cdot \dot{V} + \dot{P} \cdot V = \rho \cdot Q \cdot K \qquad (10)$$

From equation (8) the product $\rho \cdot K$ is equal to the pressure P, therefore equation 10 becomes:

$$\dot{V} = Q - \frac{\dot{P}}{P} \cdot V \qquad (11)$$

allowing to express the relation between the time variation of pulmonary volume ($\dot{V}$) with the volumetric flow rate, Q, and another term, which assumes the name of compressibility factor. Therefore, when the pulmonary volume variation is obtained by integrating the Q term over time, it is necessary to consider that the value obtained will also contain the integral of the compressibility. It will be:

$$\int Q = V + \int \frac{\dot{P}}{P} \cdot V \qquad (12)$$

The compressibility factor will be negligible only in the case of a negligible pressure rise, but this not the case of pulmonary mechanical ventilation. Therefore, if we want to obtain the lung volume variation by integration of volumetric flow rate, we will have an overestimation in inspiration ($\dot{P}$ positive) and an underestimation in expiration ($\dot{P}$ negative). This appears to be the main reason for what if we integrate the volumetric flow rate signal we obtain a slightly drifting waveform, which is supposed to be the gas volume delivered. Equation (12) let us understand that such a procedure leads to negligible errors on the calculation of the pulmonary volume only in the case of negligible pressure variation inside the lungs, that is equivalent to say that the compressibility ratio is near to zero. Therefore, only if the air compressibility contribution is supposed negligible tracheal volumetric flow rate is exactly equal to the variation of lungs geometric volume per time unit, with the advantage of being able to measure the variation of pulmonary volume through one simple measurement of volumetric flow rate in proximal airways and further integration.

CORRECTIONS DUE TO GAS TEMPERATURE AND HUMIDITY

With respect to measurements of lung volume, it should also be noted that the measured volumes must be corrected for temperature and relative humidity content of the gas inside the lungs. The delivered gas, in fact, flows from the conditions in the hospital distribution system (temperature 18-25 °C, relative humidity about 5% or negligible) to those of body of about 37 °C and saturated in term of partial pressure of water vapor (P_w), which in turn depends on temperature. In order to humidify air at 37 ºC, P_w is about 47 mmHg. The pressure at saturation can be calculated by the following equation: $P_w = 17.21\text{-}0.8937 \cdot T + 0.0460 \cdot T^2$ [mmHg], with T temperature in Celsius degrees.

The gas volume measured by the ventilator is at room temperature and dry because, generally, devices for measuring flow in ventilators are placed upstream the humidifier. Assuming gas temperature at 24 °C, P_w is about 22 mmHg. Gas at 50% relative humidity has a P_w is equal to 50% of 22 mmHg, or 11 mmHg. Since the measurements must be related to the volume occupied by gas inside the lungs, it is necessary to convert the conditions of saturated vapor pressure of the ambient temperature conditions to saturated vapor at the temperature of the human body.

$$V_B = V_A \cdot \frac{T_B \cdot (P_{atm} - P_{wA})}{T_A \cdot (P_{atm} - P_{wB})} \tag{13}$$

where V_B is the gas volume in the patient lungs, V_A the gas volume at ambient conditions, T_B is the temperature of the human body (approximately 37 °C), T_A the ambient temperature, P_{atm} the atmospheric pressure, P_{wA} and P_{wB} the partial pressures of water vapor at ambient and body conditions respectively.

From the application of equation (13), with the numerical data previously cited, a gas volume of 500ml delivered by a ventilator to a container with a perfectly mobile wall in atmospheric pressure, once realized the heating and humidifying process, would occupy a volume approximately 1.6 times higher. The above-mentioned phenomenon makes correction necessary in the case of spirometric measurements, whereas in the case of pulmonary ventilation the respiratory expansion due to heating and humidifying the gas has a smaller effect due to the contribution of lung elasticity. However, it is necessary to consider that also this correction may provide a more accurate volume estimation.

MEASUREMENTS OBTAINED FROM VENTILATORY LOOPS

Finally, it appears useful to mention some possible sources of misleading information which mainly relate to measurements associated with Pressure Volume loop (PV loop) and Flow Volume loop (FV loop). Generally speaking, every measurement obtained by two or more waveforms which should be synchronized may require the following considerations. However, the previously mentioned loops are easily obtainable from common monitoring equipment, and several studies have been conducted on their interpretation for diagnostic purposes. In particular, the PV loop may provide useful information on the shape of the static PV curve, that graphically represents the elastic characteristics of respiratory system. The recording of static PV curve for the determination of the parameters of respiratory mechanics has been subject of research and debate since the early studies on the Acute Respiratory Distress Syndrome (ARDS) by, among others, Ashbaugh *et al.* in 1967 [29].

The static PV curve of patients with ARDS, an example is reported in figure 4, shows a characteristic shape of "S", typical of this disease, with two points of major slope variation which are usually called "inflection points"[1]. Subsequent research showed that the first inflection point, at low lung volumes (Lower Inflection Point), seemed to indicate the beginning of the alveolar recruitment, whereas the second inflection point, at high lung volumes (Upper Inflection Point), indicated the beginning of alveolar over-distension. Among the first clinical applications focused on acute lung disease we find the study of Falk *et al.* in 1972 [30], where the beneficial effect of setting a Positive End Expiratory Pressure (PEEP), in the treatment of this disease, in order to avoid alveolar de-recruitment is reported. Results of this research were the indication for patients with acute lung disease, to maintain the tidal volume variation between the two “inflection points”, see figure 4, where the lower inflection point determines the appropriate level of PEEP, and the upper inflection point represents the maximum end-inspiratory pressure. Thus, the limitation of airway pressure would allow a lower risk of lung injury. The scientific community continues showing, also recently [31, 32], a considerable interest in the assessment of respiratory mechanics of artificially ventilated patients and, in particular, for the determination of PV curves as a function of time and disease stage. Therefore, it appears useful to mention some details regarding the measurement of PV loops.

As it is well known, during controlled ventilation, the mechanics of the respiratory system is substantially altered, because the pulmonary apparatus works only with positive pressure rather than, as in spontaneous ventilation, around an operating point equal to atmospheric pressure with pressure variations both positive and negative. Consequently, the mechanical properties of the respiratory system, being non-linear as the vast majority of mechanical characteristics of physiological systems, during controlled ventilation assume, in general, different values depending on the working point which is chosen the physician and set on the ventilator.

---

[1] Strictly speaking, an ‘inflection point’ is a point on a curve at which the curvature, i.e. its first derivative, changes sign. Therefore, a sigmoid, or S-shaped, curve, like the one reported in figure 4, has only one inflection point, positioned where its first derivative changes from positive to negative. However, sometimes, in literature, the inflection points are indicated as the points in which the curve first derivative changes ‘significantly’.

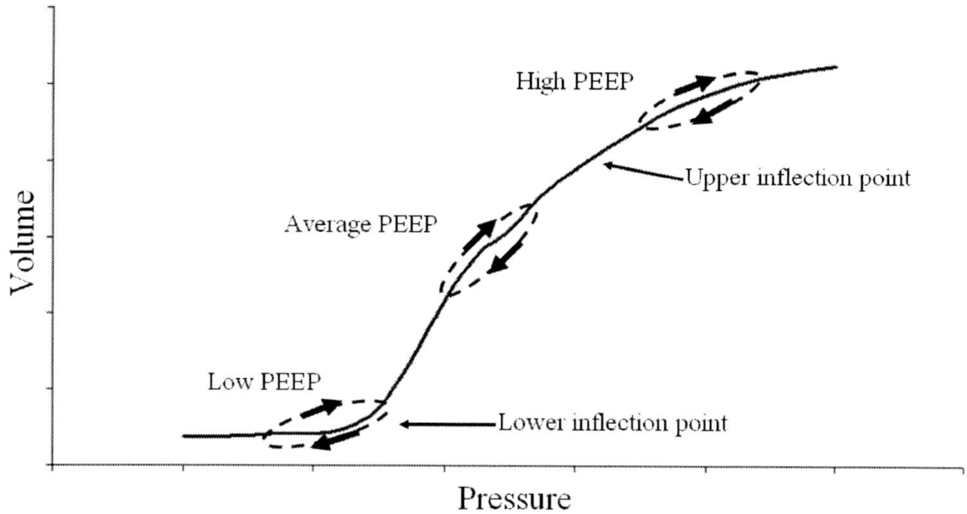

Figure 4. Representation of the static PV curve and PV loops at different PEEP values.

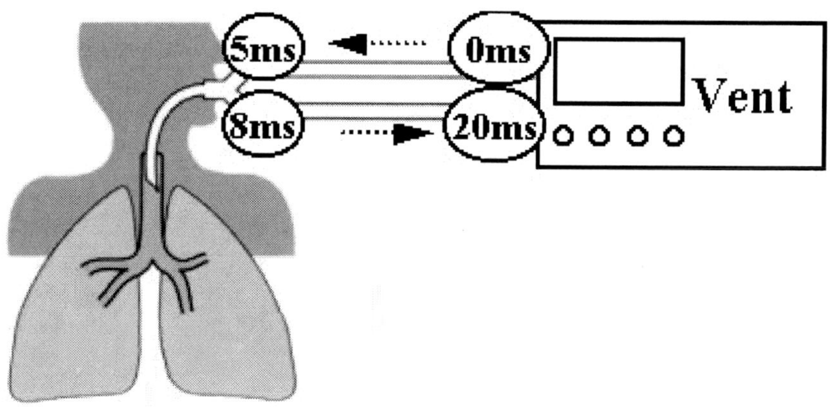

Figure 5. Typical delays of pressure increase at the beginning of forced inspiration for a standard neonatal breathing circuit.

From the observation of figure 4, the value of dynamic compliance (slope of the PV loop) is dependent not only on the dynamic variations of respiratory parameters but also on the baseline (PEEP) of alveolar pressure that, in this case, is imposed by the ventilator. Information about the static compliance can, therefore, also be obtained by considering the shape of the PV loop. For a working point near to the upper inflection point, the PV loop will show the typical *bird's beak* shape indicating a risk of alveolar over-distension. On the other hand, for a work point near the lower inflection point the PV loop will show a dynamic compliance increase with the increase of PEEP. Thus, the assessment of respiratory mechanics by PV loops appears to have the great advantage of allowing the clinician to assess respiratory mechanics without requiring the use of further devices or methods, and without disconnecting the patient from the ventilator.

The PV loop may also provide an estimate of another parameter of clinical interest: the Work of Breathing (WOB). In adults, the physiologically normal WOB is between 0.3 and

0.6 J/liter, and it can be significantly increased in the intubated patient. The clinical role of measuring WOB in respiratory diseases is still subject of debate, however, this parameter appears to provide useful information in adaptive ventilatory control [33].

From an analysis of literature, the site where the measurements of pressure and volume are performed, in order to obtain the PV loop, appears to be not indifferent and, essentially, the previously reported considerations on the patient circuit may play a significant role. In 1992, Banner *et al.* [34] reported an experimental study from which significant differences in the calculation of WOB, during spontaneous ventilation with continuous positive airway pressure, depending on the site where the pressure was measured. The patient circuit, comprising the inspiratory and expiratory limbs, can have a length of about 3 m and therefore, as it is easily seen, its pneumatic behaviour cannot, generally speaking, be neglected when performing measurements related to patient respiratory mechanics. In fact, during a controlled insufflation, the air pressure inside the circuit does not increase instantaneously along the whole extension of the breathing circuit. Typically, the pressure increase travels in the breathing circuit with a speed approximately equal to the speed of sound in air (≈340 m/s). Experimental studies have shown that, at the beginning of a controlled inspiration, the pressure increase travels from the expiratory valve back to the inspiratory one [20]. Figure 5 shows the typical delays of the pressure increase, at the very beginning of controlled inspiration, as observed experimentally, for a standard neonatal circuit (1 cm inner diameter), including humidifiers and water traps.

As it can be seen from figure 5, since pressure transducers are usually located inside the ventilator, the pressure increase, corresponding to the beginning of forced inspiration, is detected by the transducer at the beginning of the inspiratory limb about 20 ms later than the transducer at the end of the expiratory branch. We might argue, if these delays are significant from a clinical or scientific standpoint. Just to give one graphical example, in figure 6 five PV loops, obtained by the same pressure and volume waveforms during a volume controlled ventilation, are reported. The loops have been obtained with 0, 10 and 20 ms phase difference both positive and negative.

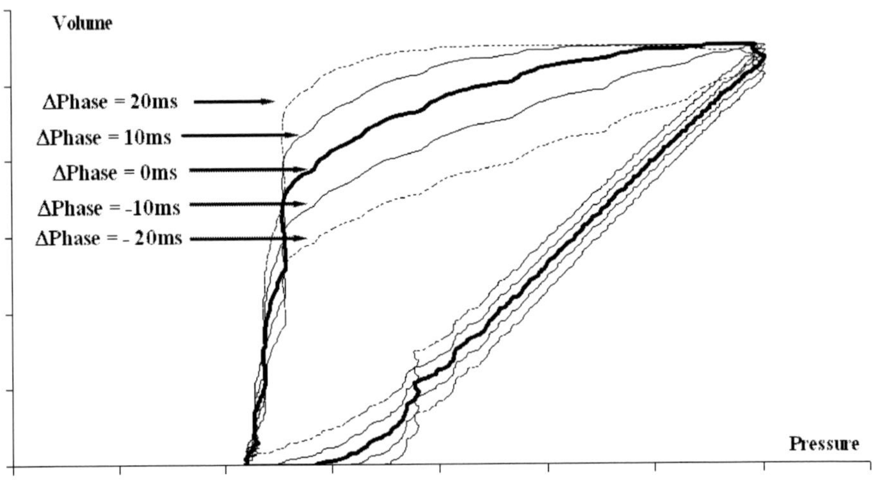

Figure 6. Effects of time delay between the pressure and volume signals on shape of the dynamic PV loop.

As it can be seen, the loop shape changes dramatically even for lags of a few tens of milliseconds. Probably, this phenomenon should be taken into account when calculating respiratory parameters, such as WOB, by integrating PV loops. For example, in some simulations, a lag of 20 ms between the pressure and volume signals may lead to an uncertainty on WOB calculation between 15-20%.

This observation may not only give an account of the experimental results obtained by Banner *et al.* [34], but also could be one reason leading some researchers to investigate the effects of the exhalation valve mechanical behaviour on the shape of the dynamic loop. The dynamic loop, in fact, shows different shapes depending on the delay between the pressure, flow and volume signals.

Conclusion

Some physical phenomena taking place in the pulmonary ventilator-patient system have been briefly described in order to address their influence on possible misleading information obtained from pulmonary monitoring units.

Generally speaking, continuous monitoring of respiratory parameters in pulmonary ventilation poses a number of daunting challenges.

Although the clinical significance may appear limited in some cases, from a scientific standpoint it is important to consider the described phenomena when evaluating the efficacy of novel ventilatory strategies and, above all, if the efficacy evaluation needs to be performed by introducing further monitoring instruments.

While hoping to have given a little contribution to the medical and engineering scientific community, further challenges in monitoring capabilities are being faced. Here we literally report from a recent paper [35] published in a major journal of biomedical engineering: "*Continuous monitoring of gas composition and volume in the complex breathing circuit environment, where changes in humidity and temperature can distort the measurement presents a number of difficulties/complications which have led to compromises and less than optimal clinical and technical solutions.*"

References

[1] Petrucci N, Iacovelli W. "Lung protective ventilation strategy for the acute respiratory distress syndrome." *Cochrane Database Syst. Rev.* 2007 Jul 18.

[2] Kilpatrick B, Slinger P. "Lung protective strategies in anaesthesia." *Br. J. Anaesth.* 2010 Dec;105 Suppl:108-16.

[3] Gattinoni L, Protti A, Caironi P, Carlesso E "Ventilator-induced lung injury: the anatomical and physiological framework." *Crit. Care Med.* 2010 Oct;38(10 Suppl):S539-48.

[4] Plataki M, Hubmayr RD. "The physical basis of ventilator-induced lung injury." *Expert Rev. Respir. Med.* 2010 Jun;4(3):373-85.

[5] Antonelli M, Azoulay E, Bonten M, Chastre J, Citerio G, Conti G, De Backer D, Lemaire F, Gerlach H, Hedenstierna G, Joannidis M, Macrae D, Mancebo J, Maggiore

SM, Mebazaa A, Preiser JC, Pugin J, Wernerman J, Zhang H "Year in review in Intensive Care Medicine 2009. Part III: mechanical ventilation, acute lung injury and respiratory distress syndrome, paediatrics, ethics, and miscellanea." *Intensive Care Med.* 36(4):567-84, 2010.

[6] Webb H H, Tierney D F. "Experimental pulmonary oedema due to intermittent positive pressure ventilation with high inflation pressures: protection by positive end-expiratory pressure" *Am. Rev. Respir. Dis*, 110, 556–565, 1974.

[7] Haake R et al. "Barotrauma. Pathophysiology, risk factors and prevention" *Chest*, 91, 608-613, 1987.

[8] Dreyfuss D, Saumon G "Barotrauma is volutrauma, but which volume is the one responsible?" *Intensive Care Med*, 18 (3), 139-141, 1992.

[9] Chao DC, Scheinhom DJ. "Barotrauma vs volutrauma" *Chest,* 109 (4), 1127-1128, 1996.

[10] Marshall J C "Critical illness is an iatrogenic disorder." *Crit. Care Med.* Oct;38(10 Suppl):S582-9, 2010.

[11] Kacmarek RM. "Ventilator-associated lung injury" *Int. Anesthesiol. Clin*, 37 (3), 47-64, 1999.

[12] Sondergaard S, Karason S, Hanson A, Nilsson K, Hojer S, Lundin S, Stenqvist O. "Direct measurement of intratracheal pressure in pediatric respiratory monitoring." *Pediatr. Res.* Mar;51(3):339-45, 2002.

[13] Kárason S, Søndergaard S, Lundin S, Wiklund J, Stenqvist O. "Direct tracheal airway pressure measurements are essential for safe and accurate dynamic monitoring of respiratory mechanics. A laboratory study" *Acta Anaesthesiol. Scand*, 45, 173-179, 2001.

[14] Cannon M L, Cornell J, Tripp-Hamel D S, Gentile M A, Hubble C L, Meliones J N and Cheifetz I M. "Tidal volumes for ventilated infants should be determined with a pneumotachometer placed at the endotracheal tube" *Am. J. Respir. Crit. Care Med.* 162, 2109-12, 2000.

[15] Castle R A, Dunne C J, Mok Q, Wade A M, Stocks J. "Accuracy of displayed values of tidal volume in the pediatric intensive care unit" *Crit. Care Med.* 30, 2566-74, 2002.

[16] Silvestri S. "The influence of flow rate on breathing circuit compliance and tidal volume delivered to patients in mechanical ventilation" *Physiol. Meas.* 27, 23-33, 2006.

[17] Heulitt M J, Thurman T L, Holt S J, Jo C H, Simpson P. "Reliability of displayed tidal volume in infants and children during dual-controlled ventilation." *Pediatr Crit. Care Med.* 10(6):661-7, 2009.

[18] Cappa P, Sciuto S A and Silvestri S. "Experimental analysis of the airway circuit effects on breathing pattern generated by neonatal pulmonary ventilators" *J. Clin. Eng.* 29 134-7, 2004.

[19] Lucangelo U, Bernabe F and Blanch L "Respiratory mechanics derived from signals in the ventilator circuit" *Respir. Care* 50, 55-65, 2005.

[20] Cappa P, Sciuto S A and Silvestri S "Experimental evaluation of errors in the measurement of respiratory parameters of the newborn performed by a continuous flow neonatal ventilator" *J. Med. Eng. Technol.* 30, 31-40, 2006.

[21] Heulitt M J, Holt S J, Thurman T L, Hall R A, Jo C H and Simpson P. "Reliability of measured tidal volume in mechanically ventilated young pigs with normal lungs" *Intensive Care Med.* 31, 1255-61, 2005.

[22] Chow L C, Vanderhal A, Raber J, Sola A. "Are tidal volume measurements in neonatal pressure controlled ventilation accurate?" *Pediatr. Pulmonol.* 34, 196-202, 2002.

[23] Neve V, Leclerc F, Noizet O, Vernoux S, Leteurtre S, Forget P, Sadik A, Riou Y. " Influence of respiratory system impedance on volume and pressure delivered at the Y piece in ventilated infants" *Pediatr. Crit. Care Med.* 4 418-25, 2003.

[24] Al-Majed S I, Thompson J E, Watson K F, Randolph A G. "Effect of lung compliance and endotracheal tube leakage on measurement of tidal volume" *Crit. Care* 8 R398-402, 2004.

[25] Khemani R G, Newth C J. "The design of future pediatric mechanical ventilation trials for acute lung injury." *Am. J. Respir. Crit. Care Med.* 15;182(12):1465-74, 2010.

[26] B G Wilson, F H Kern, I M Cheifetz "Direct measurement via an inline pneumotachometer is necessary to determine effective tidal volume" *Respir. Care*, 40, 1172, 1995.

[27] Cecchini S, Schena E, Silvestri S. "An open-loop controlled active lung simulator for preterm infants." *Med. Eng. Phys.* Jan;33(1):47-55, 2011.

[28] Ligas J R, Primiano F P. "Respiratory Mechanics" in Webster JG (Editor) Encyclopaedia of Medical Device and Instrumentation, 2550-2573, 1988.

[29] Ashbaugh D G, Bigelow D B, Petty T L, Levine B E. "Acute Respiratory Distress Syndrome in Adults" *The Lancet*, 290, 319-323, 1967.

[30] Falke K J, Pontoppidan H, Kumar A, Leith D E, Geffin B, Laver M B. "Ventilation with end-expiratory pressure in acute lung disease" *J. Clin. Invest*, 51 (9), 2315-2323, 1972.

[31] Adams A B, Cakar N, Marini J J. "Static and dynamic pressure-volume curves reflect different aspects of respiratory system mechanics in experimental acute respiratory distress syndrome." *Respir. Care*, 46 (7):686-93, 2001.

[32] Koefoed-Nielsen J, Nielsen N D, Kjaergaard A J, Larsson A. "Alveolar recruitment can be predicted from airway pressure-lung volume loops: an experimental study in a porcine acute lung injury model." *Crit. Care.* 12(1):R7, 2008.

[33] Mireles-Cabodevila E, Chatburn R L. "Work of breathing in adaptive pressure control continuous mandatory ventilation." *Respir. Care.* 54(11):1467-72, 2009.

[34] Banner M J, Kirby R R, Blanch P B. "Sitc of pressure measurement during spontaneous breathing with continuous positive airway pressure: effect on calculating imposed work of breathing" *Crit. Care Med*, 20 (4), 528-533, 1992.

[35] Jaffe M B, Orr J A. "Continuous monitoring of respiratory flow and CO(2):challenges of on-airway measurements" *Engineering in Medicine and Biology Magazine*, 29(2), 44-52, 2010.

In: Ventilation: Types, Standards and Problems
Editors: V. A. Romano and A. S. Duval, pp. 111-133
ISBN 978-1-61324-424-1

Chapter 4

INFLUENCE OF VENTILATION RATE ON RADON AND THORON PROGENY CONCENTRATIONS IN A ROOM

N. Stevanovic, V. M. Markovic and D. Nikezic*
University of Kragujevac, Faculty of Science, Kragujevac, Serbia

1. INTRODUCTION

Radon is naturally occurring radioactive element which is member of the ^{238}U radioactive decay series. The main source of radon is surrounding ground, (Eisenbud and Gasell, 1997), from which radon by diffusion and convection reaches surface and enter the atmosphere, (Fleischer, 1997). Building materials which contain higher concentration of radon parent radionuclide – radium, can be intensive sources of radon. ^{222}Rn is the most significant radioisotope because it has been estimated that more than half of the annual effective dose received by human from all sources of ionizing radiation is attributed to ^{222}Rn and its short-lived progeny (NCRP Report 93 1987; ICRP32 1981). It has the half life of 3.825 days and decays emitting alpha particle with energy of 5.49 MeV. When radon enters in dwellings it is usually uniformly distributed in the air, (Urosevic et al, 2008).

Besides radon, ^{222}Rn, thoron, ^{220}Rn, is present in our environment. Thoron is radon isotope which is member of thorium (^{232}Th) radioactive series. It decays by emission of alpha particle with energy of 6.29 MeV, with half-life of 55 s. Average concentration of ^{232}Th in soil is estimated at 25 Bq/kg, (UNESCAR Report, 2000). Mediate progeny of ^{232}Th is radium (^{224}Ra). After decay of ^{224}Ra, ^{220}Rn is formed. A typical value for ^{220}Rn concentration in pore air of deep soil is estimated to 20 kBq/m^3 (corresponding to a soil with about 25 Bq/kg ^{232}Th, porosity 50%, density 1.5g/cm^3 and emanation coefficient of 0.3), (Ramachandran, 2009). Like to radon, thoron due to chemical inertia migrates from soil into the atmosphere. In dwellings and closed spaces with poor ventilation accumulation of thoron and its progeny is possible, depending of strength of the source, (Steinhausler, 1996). In diffusion process distance that ^{220}Rn can migrate is much shorter than ^{222}Rn. However, in buildings with an ineffective barrier between soil and indoor air, the entry of ^{220}Rn could be significant.

* R. Domanovica 12. 34000 Kragujevac, Serbia.

Concentration level of thoron in dwellings depends on its emanation from soil and building materials which contains ^{232}Th, concentration of ^{232}Th itself in soil and ambient atmospheric characteristics. Problem which arises with measurements of thoron is concentration gradient which can be very pronounced (Porstendorfer 1991; Urosevic et al, 2008; Yamasaki et al, 1995).

All radon and thoron progeny are chemically reactive and can be attached to surfaces or aerosols. In recent decades, there has been a worldwide concern about the health risks from exposure to ^{222}Rn , ^{220}Rn and their progenies. Hazards from radon and thoron can be from both external and internal sources. Inhalation of indoor air is responsible for more than 80% of human exposure to ^{222}Rn and ^{220}Rn as well as to their progeny. Therefore, it is important to study the indoor behavior of ^{222}Rn and ^{220}Rn as well as their progeny from the viewpoint of dose assessment. The decay products of radon and thoron consist of a variety of nuclides and each of them may be attached to aerosols of varying sizes. Thus the airborne activity can be taken to consist of a very small 'unattached' fraction and a larger 'attached' fraction.

Decay scheme of ^{222}Rn and ^{220}Rn is presented in Figure 1.

Behavior of ^{222}Rn, ^{220}Rn and its progeny in dwellings is described by the parametric differential equations, primary given by the Jacobi, (1972). This model takes into account radioactive decay, removal by ventilation, attachment, detachment by recoil and deposition. Most authors consider radon and thoron indoor concentrations uniform inside rooms (Porstendorfer et al., 1978). Such an assumption is probably well justified and accurate for radon due to its relatively long half-life (3.825 d). However, the thoron half-life is short (56 s) and such an assumption is probably not valid. The thoron concentration in air decreases exponentially from the source. However, this might not be true if the ventilation rate is large and distribution could be more complicated (Katase et al., 1988; Yamasaki et al., 1995).

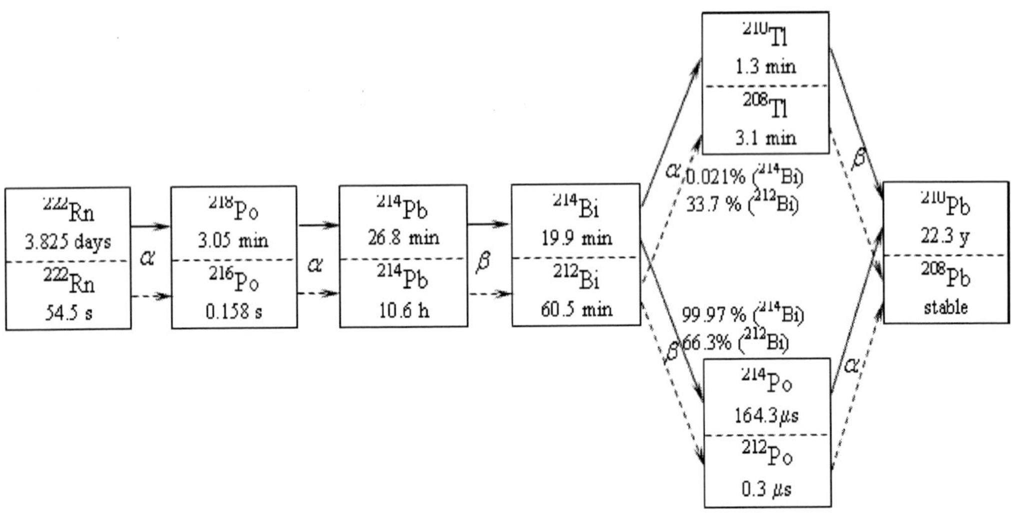

Figure 1. Radon and thoron progeny decay chain.

Thoron progeny is relatively long-lived and they could be uniformly distributed throughout a room. Such a situation makes thoron measurements problematic and determination of the health risk more complicated than in the case of radon. If a window or door is open, or some other kind of natural/artificial ventilation is present, the convection effect becomes dominant over diffusion transport of thoron atoms (Kato, 1997).

However, none of the previous studies took the airflow effect into account. Turbulence occurs in a ventilated room. It is believed that the diffusion coefficient is not uniform in general. In particular when a window/door is opened or artificial ventilation exists, the convection effect plays a more important role for mass transform than that of diffusion effect (Kato, S. 1997). Furthermore, if an activity gradient exists in a room, it is difficult to predict precisely the three-dimensional distributions from the one or two-dimensional calculations. Therefore it is essential to examine ventilation and airflow characteristics in the room in order to get full picture about its influence on radionuclide distribution in the air of dwellings.

2. Indoor Airflow and Ventilation

Ventilation is the primary mechanism for maintaining acceptable indoor air quality in modern buildings. In structures for human occupation it is important to provide a habitable, comfortable and adequately ventilated environment. Unfortunately, ventilation systems can also be responsible for the problems associated with indoor air quality like high levels of air contaminants due to insufficient flow (R. Anderson, 1988; M. Mecker, 1991). One general method to improve indoor air quality is to simply increase the ventilation rate. This approach, however, often conflicts with building energy efficiency requirements because conditioning outdoor air to the proper temperature, humidity level, and particle count for use as ventilation air consumes energy (G. Meckler, 1992). An alternate solution is to improve the efficiency of ventilation systems so that required increases in ventilation rates and energy consumption are minimal. In this case, efficiency refers to the effective delivery of properly conditioned ventilation air to building occupants rather than solely to the mechanical performance of the ventilation system. To evaluate effective delivery of air, it is important to resolve the relationships between geometric room parameters and the air flow patterns they produce.

Predicting ventilating airflow patterns would be advantageous for providing a suitable environment and the controlled removal of gaseous pollutants. Through improved understanding of ventilation mechanisms and prediction techniques, the control of temperature and relevant environmental parameters can be optimized mechanisms and prediction techniques, the control of temperature and relevant environmental parameters can be optimized.

Ventilation can be categorized as forced or natural. Forced ventilation is driven through the use of mechanical fans, for example, in air conditioning systems. This form of ventilation is capable of providing the required ventilation flow for which the system is designed and can be controlled with relative ease. Natural ventilation employs no fans, or mechanical systems, instead relying upon air flow induced by thermal gradients (buoyant flows), the action of the wind or a combination of the two mechanisms.

This part is concerned only with ventilation driven by the action of the wind. The effects of thermal gradients on natural ventilation have been found to be negligible above reasonably low wind speeds (approximately $2ms^{-1}$). Wind induced ventilation occurs due to pressure differences between openings in a building envelope and by momentum of the incoming air if the ventilation openings are large.

The most common technique currently used for predicting wind driven ventilation utilizes mean pressures measured in the positions of ventilation openings measured on sealed buildings (no openings present) and assuming a steady-state process. The prediction of wind induced ventilation through the use of mean pressure data alone has however, lead to significant discrepancies with measurements (e.g. Choiniere et al, 1992). Discrepancies become more prominent as the pressure difference between ventilation openings reduces. If the mean pressure difference across openings was zero, steady-state methods predict zero ventilation. In reality, significant air exchange has been found to take place in such cases as a result of the action of turbulent flow mechanisms.

Initially, the primary ventilation mechanisms are described followed by a more detailed description of the theory and mathematical descriptions of air exchange due to wind induced ventilation only. Finally a more in-depth discussion of the methods available for assessing ventilation and studying the phenomena involved is provided.

While ventilation due to thermal effects is described, details are provided in a qualitative sense only. This work is related to wind induced processes where thermal effects are negligible. Natural ventilation is subject to the variability of wind speed, wind direction, air temperature and opening configuration. Not only do these factors affect the rate of fresh air supply but also determine whether openings will act as an inlet or outlet for the air in any space within a building.

2.1. Deposition Velocity

The deposition mechanisms of particles on surfaces from turbulent air flow can be divided into two types (Porstendorfer , 1994): transport toward the surface, determined mainly by the eddy diffusivity; and the precipitation of aerosol particles on surfaces, a function of particle diameter and the structure and roughness of the surface. In the literature (see references in (Porstendorger, 1994)), the deposition velocity is often normalized by the friction velocity u*, where friction velocity is parameter of eddy diffusivity and depends on the air velocity profile and the roughness of the surface.

The model of particle deposition with three mechanisms of particle transport: Brownian diffusion; turbulent diffusion; and gravitational settling were developed by Lai and Nazaroff (2000). This model was extended by Zhao and Wu (2006) taking into account the mechanism of turbophoresis, defining the particle flux, J, as

$$J = -\left(\varepsilon_p + D\right)\frac{\partial C}{\partial y} - iv_s C + V_t C \qquad (1)$$

where D is the Brownian diffusivity of the particle, ε_p is the particle eddy diffusivity in the boundary layer, C is particle concentration, y is the normal distance to the surface and v_s and V_t are the settling and turbophoresis velocities, respectively. Constant i have values -1 for deposition on ceiling; 0 for vertical surface and 1 for deposition on floor.

Table 1. Characteristic values for the parameters usually used in literature

Parameters	Best estimation h^{-1}	Range of values h^{-1}
λ_v	0.55	0.1-2
λ_a	50	10-100
λ_d^u	20	5-110
λ_d^a	0.2	0.05-1.1
p_1	0.8	0.8
p_2	0	-

The expression for particle deposition velocity, developed in (Zhao and Wu, 2006), can be shortly written as Stevanovic, et al 2009b

$$v_d(u^*,d_p)=u^*\cdot\frac{F(30,u^*,d_p)}{G(u^*,d_p)} \tag{2}$$

The Eq. (2) presents deposition velocity of the particle as a function on friction velocity, u^*, and particle diameter d_p. The function F and G, with detailed descriptions were, given in (Zhao and Wu, 2006, Stevanovic, et al 2009b).

3. Behavior Radon and Thoron Progeny in a Room

3.1. Jacobi Room Model

Behavior of radon/thoron progeny in a room (or some other closed space) was described by parametric differential equations, primarily given by Jacobi (1972) which take into account radioactive decay, removal by ventilation, deposition and transition from one form to another. Parameters of the Jacobi model are: decay constants λ_i, ventilation rate λ_v, attachment rate λ_a and deposition constants of unattached and attached progeny λ_d^u , λ_d^a, resp. in s^{-1} (or h^{-1}).

In condensed form Jacobi's steady state equations are written as follows

$$f_i^u=\frac{\lambda_i(f_{i-1}^u+p_{i-1}f_{i-1}^a)}{\lambda_i+\lambda_v+\lambda_d^u+\lambda_a} \tag{3}$$

$$f_i^a=\frac{\lambda_a f_i^u+\lambda_i(1-p_{i-1})f_{i-1}^a}{\lambda_i+\lambda_v+\lambda_d^a} \tag{4}$$

where f_i is ratio of activity concentrations of ith radon progeny to radon/thoron; upper indices, ^{u,a} mean unattached and attached progeny, resp., and p_{i-1} is recoil factor defined as average probability for detachment of the recoil nuclei after radioactive decay. Characteristic values of parameters mentioned in Eqs. (3) and (4) and their ranges are given in Table 1, (Amgarou et al. 2003).

3.2. Indoor Aerosols Size Distribution with Three Modes

During the random motion, progeny cluster can meet an aerosol that is present in air and attach to it. In this way, attached progeny are formed. Detailed measurements have actually identified three modes of attached progeny. They were referred as the nucleation, accumulation and course modes (Reineking et al., 1994). Each had its own diameter $AMAD_i$ (activity median aerodynamic diameter) and geometric standard deviation, σ_{gi} (i=1, 2, 3). The notation of three modal distributions of attached radon progeny has been accepted widely. For example, the dose conversion factor and uncertainty analysis has been calculated on that basis (Marsh and Birchall, 2000; Marsh et al., 2002). Three-mode distribution concept was adopted in calculations presented in this work. Total activity size distribution of attached aerosols is obtained by summing all three distributions as

$$f_a(d) = \sum_{i=1}^{3} f_i \cdot \frac{e^{-\frac{(\ln d - \ln AMAD_i)^2}{2\cdot \ln^2(\sigma)}}}{\sqrt{2\pi}\cdot d \cdot \ln(\sigma_{gi})} \qquad (5)$$

where f_i are activity size fractions. Numerical values for $AMAD_i$, σ_{gi} and f_i used by Marsh and Birchal (2000) are: $AMAD^{nucl}$=0.05 μm, $\sigma_{g,nucl}$=2.0, f_{nucl}=0.28; $AMAD^{acc}$=0.25 μm, $\sigma_{g,acc}$=2.0 f_{acc}=0.70; $AMAD^{co}$=1.50 μm, $\sigma_{g,co}$=1.5 and f_{co}=0.02.

3.3. Detachment Probability (Recoil Factor)

Attached radon/thoron progeny diffuses through indoor air. During that motion progeny could be decayed. If the progeny is alpha emitter, new formed progeny can be detached from the carrying aerosol. The probability of detaching is detachment probability or recoil factor. Detachment probability was estimated by (Mercer, 1976) as 0.8 and this value was commonly used. Recently, (Stevanovic et al. 2004) showed that recoil factor depends on aerosol diameter, shape and its chemical structure. In addition, radon progeny can be attached to the aerosol surface, but due to coagulation of aerosols, progeny atom could be found within aerosol volume. Recoil factor is different in these two types of contamination. Here, it was assumed the following: (a) surface attachment of progeny atom to spherical aerosol, and (b) aerosol was built from quartz (SiO_2). With these assumptions the recoil factor was distributed in the range between 0.5 and 1. The distribution of recoil factor as a function of aerosol diameter was created.

In the paper from (Stevanovic et al. 2004) was shown that probability ^{214}Pb of detachment was equal to 1 if diameter of aerosol from SiO_2 is smaller than 0.047 μm. For larger aerosols detachment probability, as function on aerosol diameter, can be presented as

$P_d(d_a) = 0.44 + 0.07e^{-0.012 \cdot d_a} + 0.54e^{-10.22 \cdot d_a} + 0.06e^{-1.32 \cdot d_a}$. Detachment was occurred if sampled random number, γ, satisfies condition $\gamma \le P_d(d_a)$. The aerosols are usually contaminated on its surface. In most cases (>50%) newly formed ^{214}Pb atoms emit from aerosol surface and detachment can be treated as immediately.

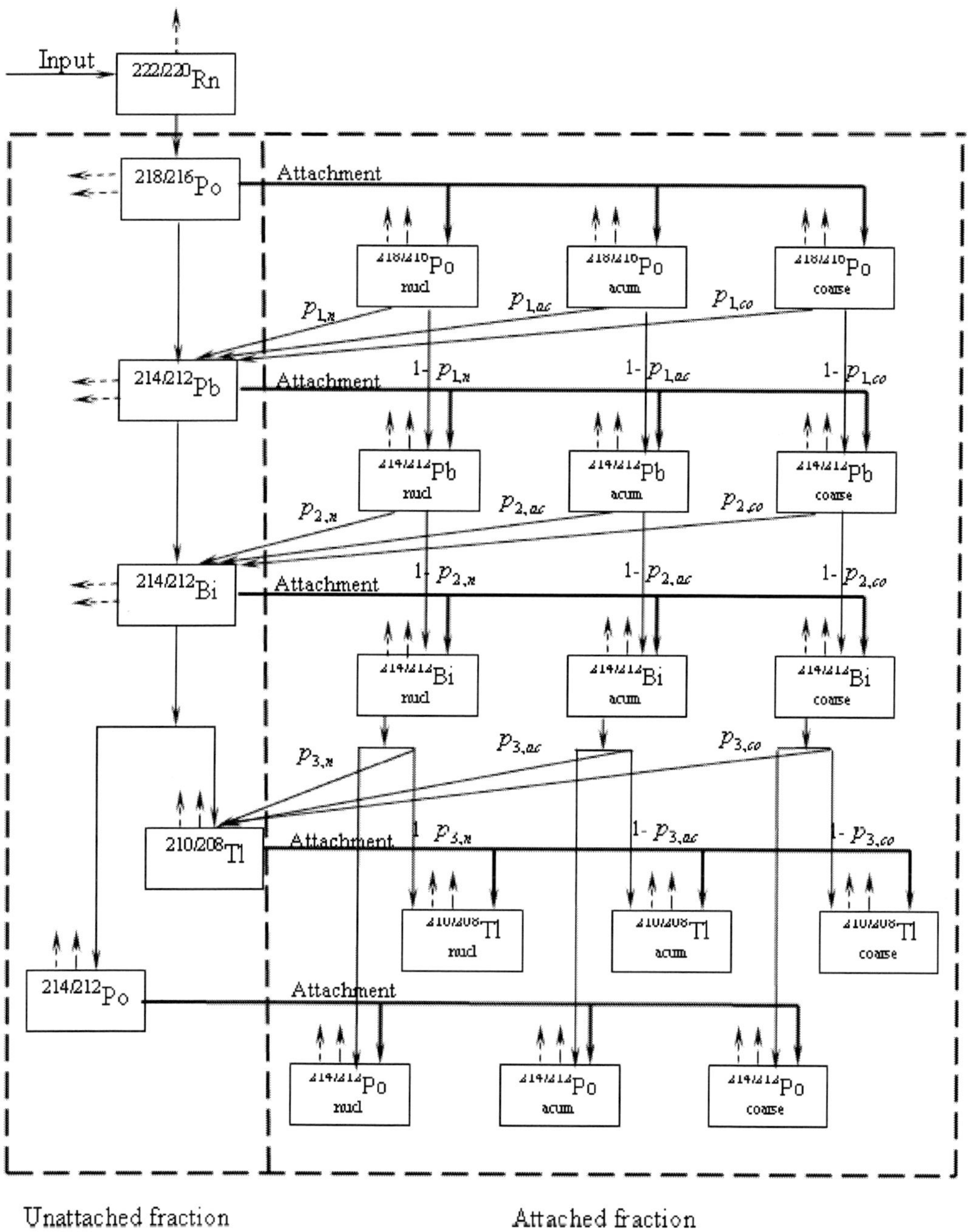

Figure 2. Behavior of radon/thoron progeny in a room with three modal aerosols distribution.

3.4. Extended Jacobi Room Model

In the paper (Nikezic and Stevanovic, 2004., 2007) a new system of steady state equations which describes concentrations of radon/thoron progeny in a room was developed by taking into account three modal distribution of attached fraction. It is improved and extended Jacobi room model according to three modal aerosols distribution. The block diagram of indoor radon/thoron progeny behavior was presented in Figure 2. Two boxes were presented in the Figure 2, left box is unattached fraction; right box is attached radon/thoron progeny. The processes of the progeny behavior were presented by lines: dashed lines are progeny ventilation from the room; dash-dot lines are progeny deposition on the surface room; solid lines are decays; thick solid lines are attachment or detachment of the progeny.

According to Figure 2, the mathematical Jacobi room model was extended in (Nikezic and Stevanovic, 2004., 2007). First two equations that describe radon/thoron and unattached $^{218/216}$Po concentrations are the same as in original Jacobi model:

$$\frac{dN_0}{dt} = S - \lambda_0 N_0 - \lambda_v N_0 \tag{6}$$

where N_0 is number concentration of ^{222}Rn atoms·m^{-3} and S is the strength of the radon source in atoms·m^{-3}·s^{-1}, λ_0 decay constant of $^{222/220}$Rn, λ_v ventilation removal rate. By decay of $^{222/220}$Rn an unattached atom of $^{218/216}$Po is formed in air which number concentration in air is described by

$$\frac{dN_1^u}{dt} = \lambda_0 N_0 - N_1^u\left(\lambda_1 + \lambda_v + \lambda_d^u + \lambda_a\right) \tag{7}$$

where N_1^u is number concentration of unattached $^{218/216}$Po in atom·m^{-3}. The first term in Eq.(7), $\lambda_0 N_0$, represents the formation of $^{218/216}$Po by decay of $^{222/220}$Rn. Unattached atoms can be attached to the aerosols presented in air with the attachment rate λ_a, or can be removed by deposition, λ_d^u, ventilation, λ_v or decay λ_1.

The following three equations describe $^{218/216}$Po concentration in three modes of attached form:

$$\frac{dN_1^{nucl}}{dt} = \lambda_a^{nucl} N_1^u - N_1^{nucl}\left(\lambda_1 + \lambda_v + \lambda_d^{nucl}\right) \tag{8}$$

$$\frac{dN_1^{ac}}{dt} = \lambda_a^{ac} N_1^u - N_1^{ac}\left(\lambda_1 + \lambda_v + \lambda_d^{ac}\right) \tag{9}$$

$$\frac{dN_1^{co}}{dt} = \lambda_a^{co} N_1^u - N_1^{co}\left(\lambda_1 + \lambda_v + \lambda_d^{co}\right) \tag{10}$$

where N_1^{nucl}, N_1^{ac} and N_1^{co} are atom number concentrations of $^{218/216}Po$ in nucleation, accumulation and coarse mode. The first terms in right side of Eqs.(8-10) represent attachment of unattached $^{218/216}Po$ to particular mode. $\lambda_a^{nucl}, \lambda_a^{ac}$ *and* λ_a^{co} are attachment rate to nucleation, accumulation and coarse mode. Second terms in right sides of Eqs.(8-10) are losing of $^{218/216}Po$ atoms due to the decay, ventilation, and deposition. Different deposition rates, λ_d^{nucl}, λ_d^{ac} and λ_d^{co} are assumed in three modes. This way of equations development can be continued, and is given as follows:

$$\frac{dN_2^u}{dt} = \lambda_1 N_1^u + \lambda_1\left(p_{1,n} N_1^{nucl} + p_{1,ac} N_1^{ac} + p_{1,co} N_1^{co}\right) - N_2^u\left(\lambda_2 + \lambda_v + \lambda_d^u + \lambda_a\right) \tag{11}$$

$$\frac{dN_2^{nucl}}{dt} = \lambda_a^{nucl} N_2^u + \left(1 - p_{1,n}\right)\lambda_1 N_1^{nucl} - N_2^{nucl}\left(\lambda_2 + \lambda_v + \lambda_d^{nucl}\right) \tag{12}$$

$$\frac{dN_2^{ac}}{dt} = \lambda_a^{ac} N_2^u + \left(1 - p_{1,ac}\right)\lambda_1 N_1^{ac} - N_2^{ac}\left(\lambda_2 + \lambda_v + \lambda_d^{ac}\right) \tag{13}$$

$$\frac{dN_2^{co}}{dt} = \lambda_a^{co} N_2^u + \left(1 - p_{1,co}\right)\lambda_1 N_1^{co} - N_2^{co}\left(\lambda_2 + \lambda_v + \lambda_d^{co}\right) \tag{14}$$

$$\frac{dN_3^u}{dt} = \lambda_2 N_2^u + \lambda_2\left(p_{2,n} N_2^{nucl} + p_{2,ac} N_2^{ac} + p_{2,co} N_2^{co}\right) - N_3^u\left(\lambda_3 + \lambda_v + \lambda_d^u + \lambda_a\right) \tag{15}$$

$$\frac{dN_3^{nucl}}{dt} = \lambda_A^{nucl} N_3^u + \left(1 - p_{2,n}\right)\lambda_2 N_2^{nucl} - N_3^{nucl}\left(\lambda_3 + \lambda_v + \lambda_d^{nucl}\right) \tag{16}$$

$$\frac{dN_3^{ac}}{dt} = \lambda_a^{ac} N_3^u + \left(1 - p_{2,co}\right)\lambda_2 N_2^{ac} - N_3^{ac}\left(\lambda_3 + \lambda_v + \lambda_d^{ac}\right) \tag{17}$$

$$\frac{dN_3^{co}}{dt} = \lambda_a^{co} N_3^u + \left(1 - p_{2,co}\right)\lambda_2 N_2^{co} - N_3^{co}\left(\lambda_3 + \lambda_v + \lambda_d^{co}\right) \tag{18}$$

In above equations, $p_{1,n,ac,co}$ are recoil factors after alpha decay of $^{218/216}Po$ in corresponding modes, and $p_{2,n,ac,co}$ are recoil factors after beta decay of $^{214/212}Pb$, that are taken to be zero. These equations are for $^{218/216}Po$, $^{214/212}Pb$ and $^{214/212}Bi$ progeny, respectively.

Bifurcation in $^{222/220}Rn$ chain occurs after decay of $^{214/212}Bi$, when two progeny, $^{210/208}Tl$ or $^{214/212}Po$ are formed with different probabilities as shown in Figure 1. ^{210}Tl is formed after alpha decay of ^{214}Bi with probability p_b=0.021% , while ^{208}Tl is formed after alpha decay of

^{212}Bi with probability p_b=33.7%. Due to the recoil, new formed progeny can be detached from the carrying aerosol. Detachment probabilities are different in each of modes and they are denoted as $p_{3,n}$, $p_{3,ac}$, $p_{3,co}$. The following equations describe $^{210/208}$Tl number concentrations $N_4^{u;nucl;ac;co}$ in unattached fraction and the three modes.

$$\frac{dN_4^u}{dt} = p_b\lambda_3 N_3^u + p_b\lambda_3\left(p_{3,n}N_3^{nucl} + p_{3,ac}N_3^{ac} + p_{3,co}N_3^{co}\right) - N_4^u\left(\lambda_4 + \lambda_v + \lambda_d^u + \lambda_a\right) \tag{19}$$

$$\frac{dN_4^{nucl}}{dt} = \lambda_a^{nucl}N_4^u + p_b\left(1 - p_{2,n}\right)\lambda_3 N_3^{nucl} - N_4^{nucl}\left(\lambda_4 + \lambda_v + \lambda_d^{nucl}\right) \tag{20}$$

$$\frac{dN^{ac}}{dt} = \lambda_a^{ac}N_4^u + p_b\left(1 - p_{2,ac}\right)\lambda_3 N_3^{ac} - N_4^{ac}\left(\lambda_4 + \lambda_v + \lambda_d^{ac}\right) \tag{21}$$

$$\frac{dN^{co}}{dt} = \lambda_a^{co}N_4^u + p_b\left(1 - p_{2,co}\right)\lambda_3 N_3^{co} - N_4^{co}\left(\lambda_4 + \lambda_v + \lambda_d^{co}\right) \tag{22}$$

Branching ratio of p_b is taken into account in the previous equations. Finally, the equations for $^{214/212}$Po which is formed from $^{214/212}$Bi, with branching ratio of $1\text{-}p_b$, are given below

$$\frac{dN_5^u}{dt} = \left(1 - p_b\right)\lambda_3 N_3^u - N_5^u\left(\lambda_5 + \lambda_v + \lambda_d^u + \lambda_a\right) \tag{23}$$

$$\frac{dN_5^{nucl}}{dt} = \lambda_a^{nucl}N_5^u + \left(1 - p_b\right)\lambda_3 N_3^{nucl} - N_5^{nucl}\left(\lambda_5 + \lambda_v + \lambda_d^{nucl}\right) \tag{24}$$

$$\frac{dN_5^{ac}}{dt} = \lambda_a^{ac}N_5^u + \left(1 - p_b\right)\lambda_3 N_3^{ac} - N_5^{ac}\left(\lambda_5 + \lambda_v + \lambda_d^{ac}\right) \tag{25}$$

$$\frac{dN_5^{co}}{dt} = \lambda_a^{co}N_5^u + \left(1 - p_b\right)\lambda_3 N_3^{co} - N_5^{co}\left(\lambda_5 + \lambda_v + \lambda_d^{co}\right) \tag{26}$$

The characteristic values of Jacobi room model parameters are given in Table 2 (Nikezic and Stevanovic, 2004; 2007).

Concentration of radon/thoron progeny were calculated according to extended Jacobi room model and their fractions were presented as a function of ventilation rate on the Figs. 3-6.

It can be seen that unattached fractions of radon/thoron progeny are almost constant functions of ventilation rate. The fractions of free radon progeny are smaller than 20% for examined range of ventilation rate.

The Figures 4 and 6 present accumulation mode of attached fraction. The attached progeny fractions strongly decrease with ventilation rate.

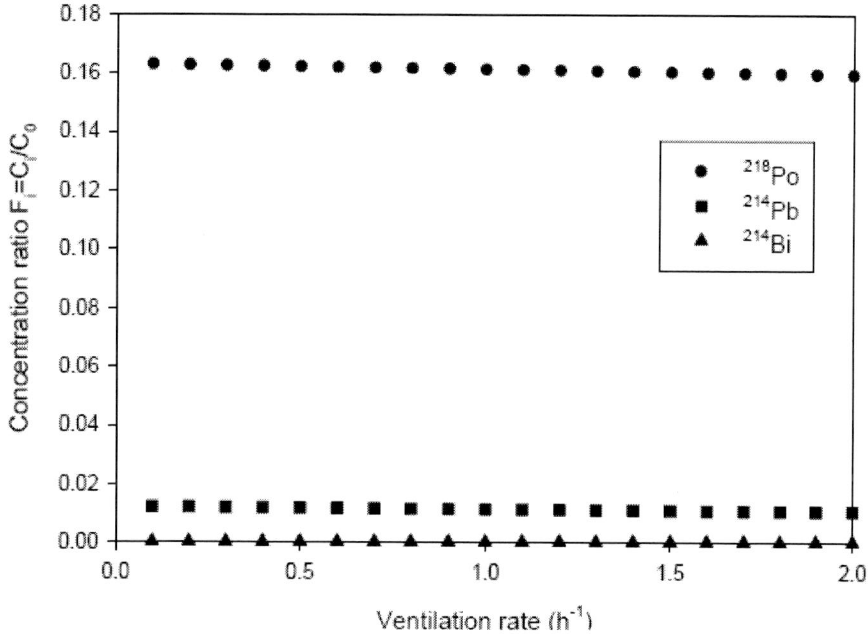

Figure 3. The fractions of unattached radon progeny as a function on ventilation rate.

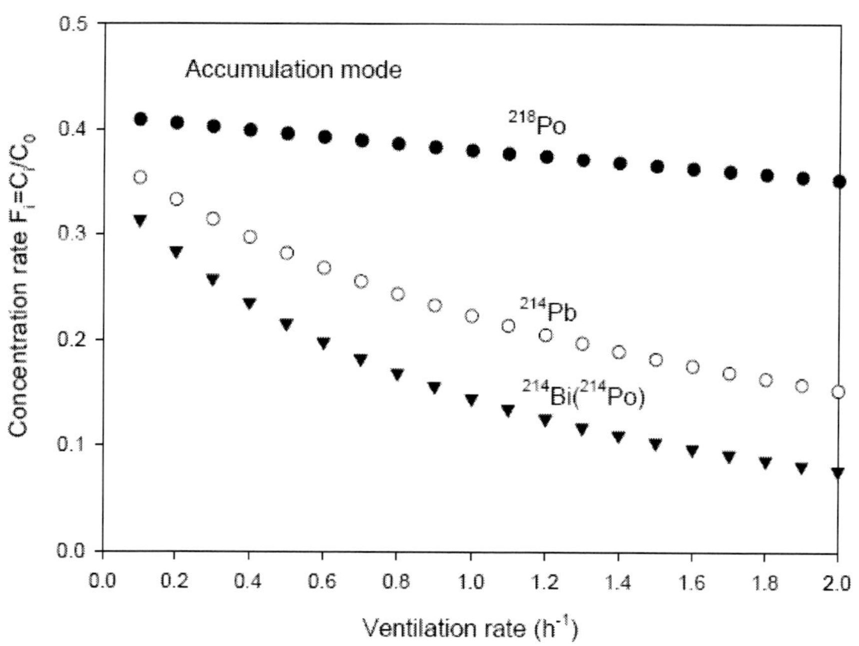

Figure 4. The fractions of attached accumulation mode of radon progeny as a function on ventilation rate.

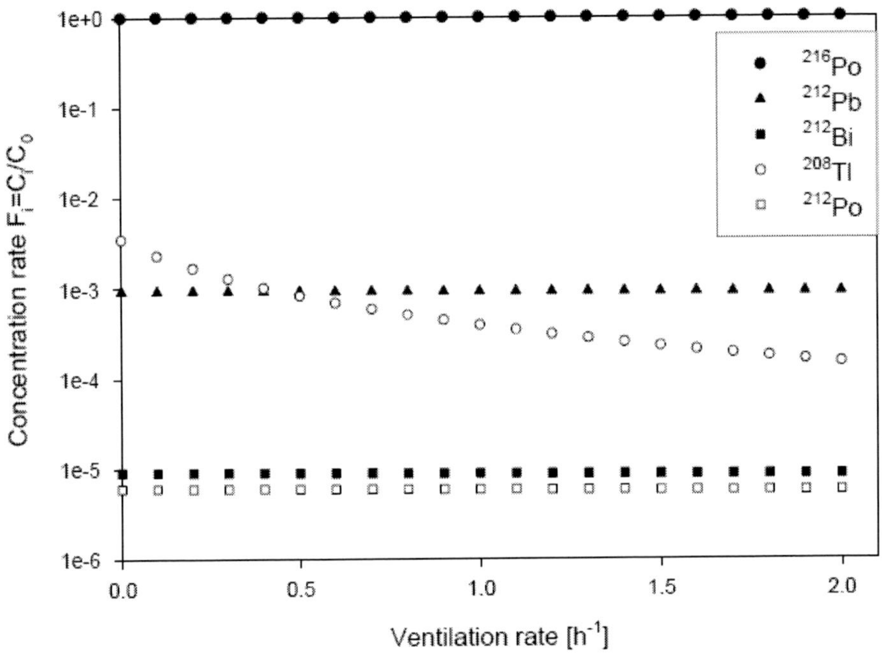

Figure 5. The ratio of unattached progeny activity concentrations to thoron concentration as a function on ventilation rate.

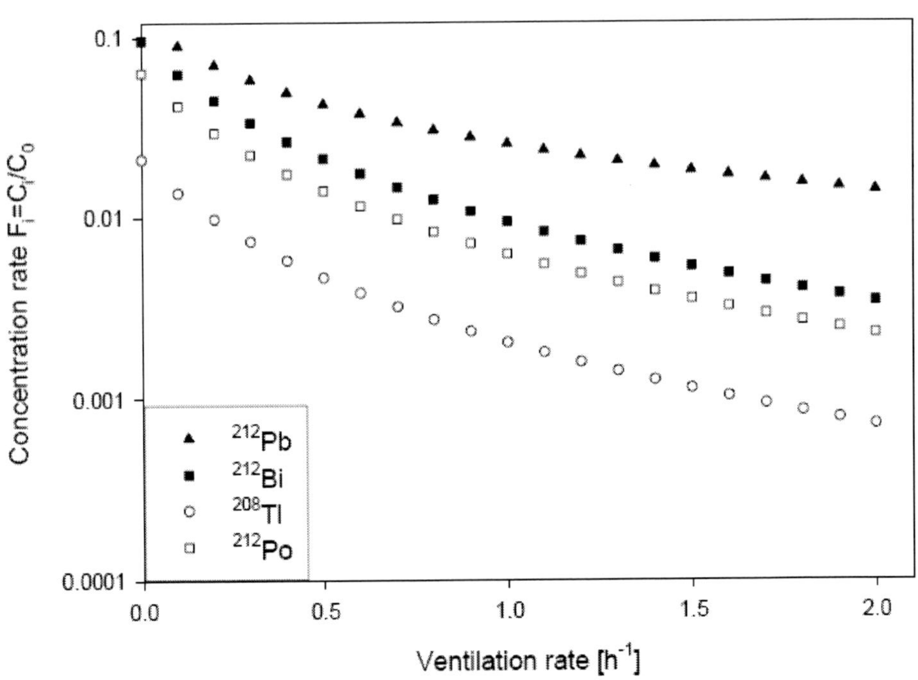

Figure 6. The ratio of attached progeny activity concentrations in accumulation mode to thoron activity concentration as a function on ventilation rate.

Table 2. Values of parameters used as input in Jacobi room model

Parameters	Values
λ_v	0.55 h^{-1}
λ_a	50 h^{-1}
λ_d^u	20 h^{-1}
λ_d^{nucl}	1.2 h^{-1}
λ_d^{ac}	0.2 h^{-1}
λ_d^{ac}	0.2 h^{-1}
$p_{1,nuc}^{1}$	0.95
$p_{1,ac}$	0.61
$p_{1,co}$	0.52
$p_2^{nucl,acc,co}$	0.
$p_{3\ nucl}$	0.87
$p_{3\ ac}$	0.60
$p_{3\ co}$	0.51

Table 3. Ranges of Jacobi model parameters estimated by Brownian motion

Parameters of Jacobi model		Amgarou et al, 2003	Nikolopoulos and Vogiannis, 2007		Stevanovic et al., 2009a
			Eftalou spa	Thermi spa	
Attachment rate	^{218}Po	5-500	50-52	200-203	60-170
	^{214}Pb		105-112	90-96	41-120
	^{214}Bi		0.5-0.8	0.8-0.9	43-122
Deposition rate of unattached progeny	^{218}Po	5-110	102-103	169-171	39-47
	^{214}Pb		112-120	98-105	30-36
	^{214}Bi		0.6-0.7	0.5-0.6	31-37
Deposition rate of attached progeny	^{218}Po	0.05-1.1	4.9-5.0	0.9-1.1	0.014-0.021
	^{214}Pb		0.9-1.0	0.25-0.26	0.005-0.008
	^{214}Bi		3.7-4.0	1.8-2.1	0.006-0.009

4. Parameters of Jacobi Room Model without Airflow

To determine the concentration of radon/thoron and their progenies in a room it is necessary to know the values of Jacobi room model parameters. Their values depend on conditions in the given indoor room, such as temperature, air motion, concentration and distribution of aerosols, ratio of surface to room volume, S/V, etc. If air motion in the room is neglected, the Jacobi room parameters can be determined by modeling behavior aerosols particle and progeny by Brownian motion.

This is justify if there is not temperature gradient in the room i.e. temperature field is uniform. In that case there is not larger air movement into the room, so the indoor airflow can be neglected and ventilation rate is equal to $\lambda_v=0$.

Determination of Jacobi room parameters with these assumptions was performed by (Stevanovic et al., 2009a). The room with common dimensions of 4 x 3 x 2.5 m^3 was considered. Deposition rate and attachment rate as a function of aerosol concentrations were considered while aerosol concentration was varied in range $(1\text{-}10)\cdot 10^3$ cm^{-3}

The values of Jacobi model parameters obtained in (Stevanovic et al 2009a) have been shown in the Table 3 and compared with values shown in (Amgarou et al, 2003, Nikolopoulos and Vogiannis, 2007).

In the paper (Nikolopoulos and Vogiannis, 2007) the parameters were determined from activity concentration of radon progeny measured in thermal spas Eftalou and Thermi.

The parameters taken from (Amgarou et al, 2003) are same for all radon progeny, while in (Nikolopoulos and Vogiannis, 2007, Stevanovic et al, 2009a) the values have been estimated for each of three radon progeny, ^{218}Po, ^{214}Pb and ^{214}Bi. It could be seen that ranges of the corresponding parameters values in these works are same order, but different. The values of deposition rates of unattached and attached progeny given in (Nikolopoulos and Vogiannis, 2007) are overestimated. It is due to larger humidity in examined thermal spas and it enlarges deposition of progeny. The deposition rate of attached progeny, estimated in (Stevanovic et al 2009a), are smaller then another. It is due to two reasons: first, the aerosols slower diffuse through the room than free progeny clusters; the second, the airflow into the room, which enlarges deposition on the surface, was not taken into account in (Stevanovic et al 2009a).

5. Parameters of Jacobi Room Model with Indoor Airflow

Ventilation in some rooms additionally influences airflow distribution and depends on wind speed and the configuration of apertures in walls, floor and ceiling. The movement of the particles in the room as well as radon/thoron progeny will be different in respect to that if there is not airflow. It does mean that indoor airflow additionally influences deposition rate of progeny as well as their attachment. The Jacobi room parameters if there is indoor airflow were considered in this section.

5.1. Ventilation Rate in Turbulent Indoor Airflow

Ventilation rate, λ_v, of radon and thoron progeny is defined as a fraction of indoor air which is exchanged per unit time Porstendoirfer, 1994

$$\lambda_v[h^{-1}] = \frac{V_{ex}}{V} \tag{27}$$

where V_{ex} (m^3/h) is air volume which is exchanged between indoor and outdoor in unit time and V is room volume (m^3).

If U [cm/s] is inlet air speed in the room, S_e is surface where air enter in the room, V is room volume, then the ventilation rate is

$$\lambda_v[h^{-1}] = \frac{U \cdot S_e \cdot 3600}{V} \tag{28}$$

Vertical profile of wind speed in urban environment is described by log-law model (Straw, 2000) as

$$U(z) = \frac{u^*}{k} \ln\left(\frac{z + z_0}{z_0}\right) \tag{29}$$

where, k is constant and has the value of k = 0.4, z is height and z_0 is aerodynamic surface roughness and for cities and village has value z_0 = 1 m (Straw, 2000).

By combining the Eqs. 8 and 9, the following can be written

$$\lambda_v = \frac{S_e \cdot 3600}{V} \cdot \frac{u^*}{k} \ln\left(\frac{z + z_0}{z_0}\right) \tag{30}$$

If inlet area is 20 x 20 cm^2 on the height z = 1.5 m (Zhuo, 2001), then friction velocity due to ventilation is equal to

$$u^* = \frac{V}{3.3 \cdot 10^6} \lambda_v \tag{31}$$

For the room with usual dimensions 400 x 300 x 250 cm^3 the increment of friction velocity (in cm/s) due to room ventilation, with ventilation rate λ_v, (in h^{-1}) is equal to Stevanovic et al, 2009b.

$$u^* = 9.09 \cdot \lambda_v \tag{32}$$

The increment of friction velocity, given by Eq. (31), presents contribution to total friction velocity in the room due to ventilation. This contribution will influence on the deposition velocities as well as deposition rates of the unattached and attached fractions.

5.2. Deposition Rate of Unattached Fraction in Turbulent Indoor Airflow

Newly formed radon progeny atoms are usually positively charged particles. During their motion they collide with molecules of water vapor and form clusters. Such clusters are known as unattached progeny. Dimensions of clusters are about 1 nm and their size distribution is described by one modal log - normal distribution, $f_u(d)$ as

$$f_u(d) = \frac{e^{-\frac{(\ln d - \ln AMTD)^2}{2\cdot \ln^2(\sigma)}}}{\sqrt{2\pi}\cdot d\cdot \ln(\sigma)} \tag{33}$$

where AMTD=0.9 nm and σ=1.3 (ICRP66, 1994).

The average deposition velocity for unattached progeny atoms, as a function of friction velocity, was calculated as (Stevanovic et al 2009b)

$$\bar{v}_d^u(u*) = \frac{\int v_d(u*, d_p)\cdot f_u(d_p)dd_p}{\int f_u(d_p)dd_p} \tag{34}$$

where $v_d(u*,d_p)$ is deposited velocity of the particle in indoor airflow described in Section 2.1. It can be seen from Eq.(2) that average deposition rate of unattached fraction, $\bar{v}_d^u(u*)$, is function on friction velocity, u*. This function was presented on the Figure 7 and it could be seen that deposition velocities of unattached fraction linearly increase with friction velocity.

In order to determine correlation between deposition velocity and friction velocity, curve on Figure 7 was fitted using iteration method of changing initial constants in probe linear function and following expression was obtained Stevanovic et al 2009b

$$\bar{v}_d^u = 0.042\cdot u^* \tag{35}$$

If deposition velocity of unattached radon progeny, v_d^u [cm/s], is known, then deposition rate, λ_d^u [h^{-1}], is given as Stevanovic et al, 2009b

$$\lambda_d^u(u*) = \frac{S}{V}\cdot 3600\cdot \bar{v}_d^u(u*) \tag{36}$$

where *S* and *V* are surface and volume of the room, respectively. Because unattached progeny clusters have very little diameters, influence of gravitational force could be neglected, and it was assumed that deposition velocity has same value at all walls of the room.

Deposition rates can be determined from the known deposition velocities. For that purpose the room with usual dimensions 400 x 300 x 250 cm^3 was considered.

By using deposition velocity of unattached fraction, v_d^u, given in Eqs.(35), deposition rate of unattached fraction, λ_d^u, calculated from Eq. (36) becomes

$$\lambda_d^u = 2.97\cdot u* \tag{37}$$

Ventilation of the room increases friction velocity. The increment of friction velocity due to ventilation can be determined from Eq.(32) as $\Delta u^* = 9.09 \cdot \lambda_v$ and Eq.(37) becomes

$$\lambda_d^u = 2.97 \cdot (u^* + 9.09 \cdot \lambda_v) \tag{38}$$

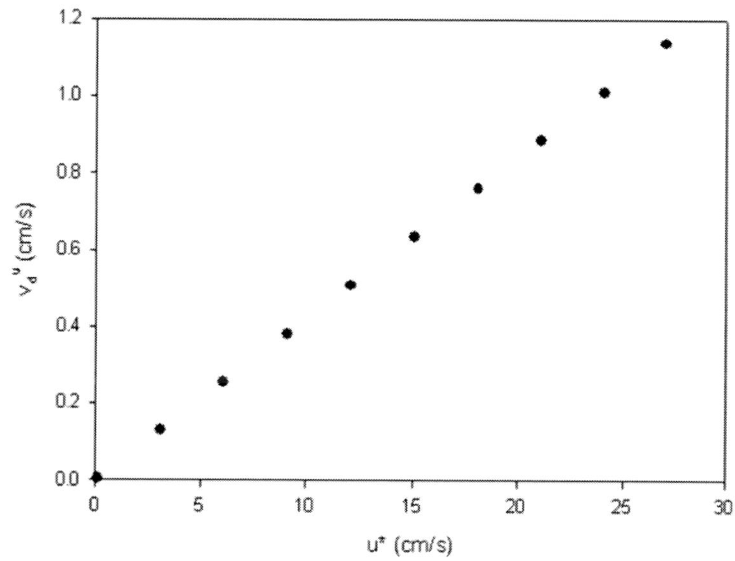

Figure 7. Deposition velocity of unattached fraction as function on friction velocity.

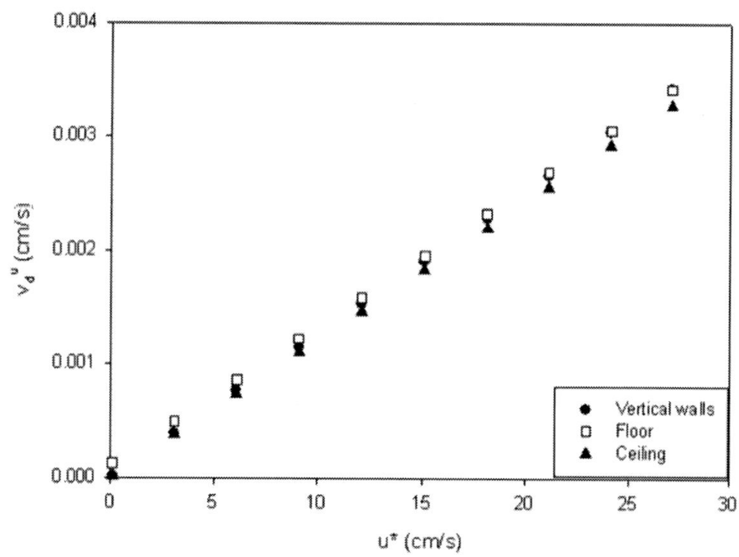

Figure 8. Deposition velocities of attached fraction on vertical wall and floor as function on friction velocity.

5.3. Deposition Rate of Attached Fraction in Turbulent Indoor Airflow

During the random motion, progeny cluster can meet an aerosol that is present in air and attach to it. In this way, attached progeny are formed. Distribution of attached radon/thoron progeny was given by Eq (5).

The average deposition velocity for attached progeny atoms, as a function on friction velocity, was calculated as

$$\bar{v}_d^a(u^*) = \frac{\int v_d(u^*, d_p) \cdot f_a(d_p) dd_p}{\int f_a(d_p) dd_p} \qquad (39)$$

Indoor aerosols carrying attached progeny have larger diameters and gravitational force has significant influence on deposition velocity at floor and ceiling. Consequently, deposition velocity on the floor is larger than that on the vertical wall. Hence, deposition rate of attached fraction is defined as

$$\lambda_d^a(u^*) = 3600 \cdot \frac{\bar{v}_{d\,wall}^a(u^*) \cdot S_{wall} + \bar{v}_{d\,floor}^a(u^*) \cdot S_{floor} + \bar{v}_{d\,ceiling}^a(u^*) \cdot S_{ceiling}}{V} \qquad (40)$$

where S_{wall}, S_{floor} and $S_{ceiling}$ are surface of the vertical walls, floor and ceiling, respectively. $v_d{}^a{}_{wall}$, $v_d{}^a{}_{floor}$ and $v_d{}^a{}_{ceiling}$ are corresponding deposition velocities.

The aerosols move through the room air slower than unattached radon progeny. Therefore, progeny attached to the aerosols have lower velocity than unattached once. Deposition velocity of attached progeny is smaller than the deposition velocity of unattached progeny. Deposition velocities of attached fraction to vertical walls, floor and ceiling in examined range of friction velocity were presented in Figure 8.

Deposition velocity on the floor is larger than on the vertical walls and ceiling due to gravitational settling. For examined friction velocity, deposition velocities of attached fraction have values up to 0.004 cm/s.

In order to determine correlation between deposition velocity of attached fraction and friction velocity, curves on Fig 8 were fitted and it have been obtained following equations Stevanovic et al 2009b

$$\begin{aligned}
\bar{v}_{d\,wall}^a &= 1.26 \cdot 10^{-4} \cdot u^* \\
\bar{v}_{d\,floor}^a &= 1.22 \cdot 10^{-4} \cdot u^* + 4.33 \cdot 10^{-4} \\
\bar{v}_{d\,ceiling}^a &= 1.21 \cdot 10^{-4} \cdot u^*
\end{aligned} \qquad (41)$$

Deposition velocity of unattached fraction is larger than that of attached fraction for about two orders of magnitude. The main reasons for that is diffusion coefficient, which is much larger for unattached than for attached fraction. Hence, the unattached progeny diffuse faster through the room. For examined friction velocity of u* = (0.1-30) cm/s, the deposition velocity of unattached fraction varies from 0.004 up to 1.2 cm/s.

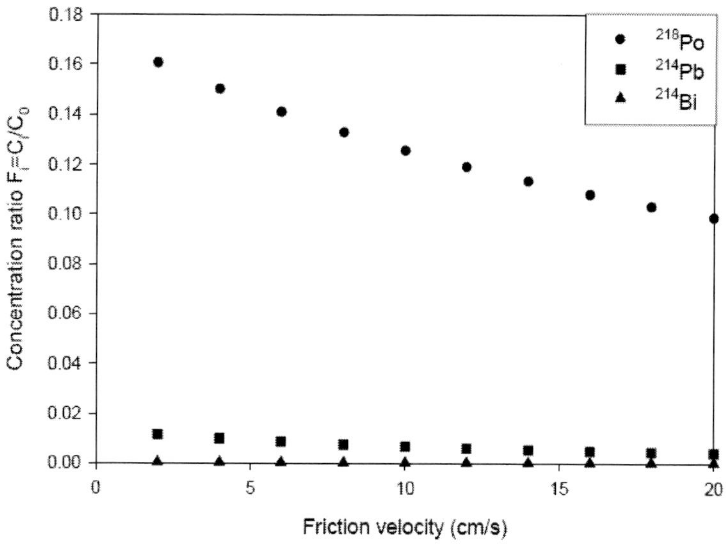

Figure 9. Concentration ratios of unattached radon progeny as a function on friction velocity.

Deposition rate for attached progeny for room with usual dimensions 400 x 300 x 250 cm^3 as a function on friction velocity and ventilation rate has a form (Stevanovic et al 2009b)

$$\lambda_d^a = 8.8 \cdot 10^{-3} \cdot (u^* + 9.09 \cdot \lambda_v) + 6.2 \cdot 10^{-3} \tag{42}$$

5.4. Attachment Rate of Attached Fraction in Turbulent Indoor Airflow

In pervious sections, by using the particle deposition model for turbulent indoor airflow and assuming that airflow is present within the room, the deposition rates of unattached and attached radon/thoron progeny were determined as functions on friction velocity (Stevanovic et al., 2009b). The room ventilation increases air movement and friction velocity. Hence, increasing of ventilation rate the deposition rates of unattached and attached progeny also increase and relationships between these three parameters were found (Stevanovic et al., 2009b). As consequence, these parameters cannot be treated as independent between themselves.

On the other side, attachment rate and deposition rate depend on aerosols concentration and their size distribution. Thus, these two parameters are not independent. Relationship between them was found (Stevanovic et al., 2010) and influence of ambient aerosols characteristics on these parameters as well as ventilation rate was investigated in this section.

Stevanovic, et al. (2010) found relationship between attachment rate, ventilation rate and deposition rate of attached fraction as a function on friction velocity

$$\lambda_d^a = 8.58 \cdot 10^{13} \cdot \left(\frac{\lambda_a}{N_0}\right)^{1.9} + 9.8 \cdot 10^{-9} \cdot \left(\frac{\lambda_a}{N_0}\right)^{-0.68} \cdot (u^* + 9.09 \cdot \lambda_v) \tag{43}$$

6. Radon/Thoron Progeny Concentration in Turbulent Indoor Airflow

The fractions of the radon/thoron progeny concentrations as a function on friction velocity, u*, were calculated by using set of Eqs. (6-26) of extended Jacobi room model. Deposition rates for unattached and attached progeny atoms were calculated by Eq. (38) and (43). It was considered that aerosols concentration is $N_0=10^{10}$ m^{-3} while attachment and ventilation rate is $\lambda_a=50$ h^{-1} and $\lambda_v=0.55$ h^{-1}, respectively. Some from results are given in Figs. 9-11.

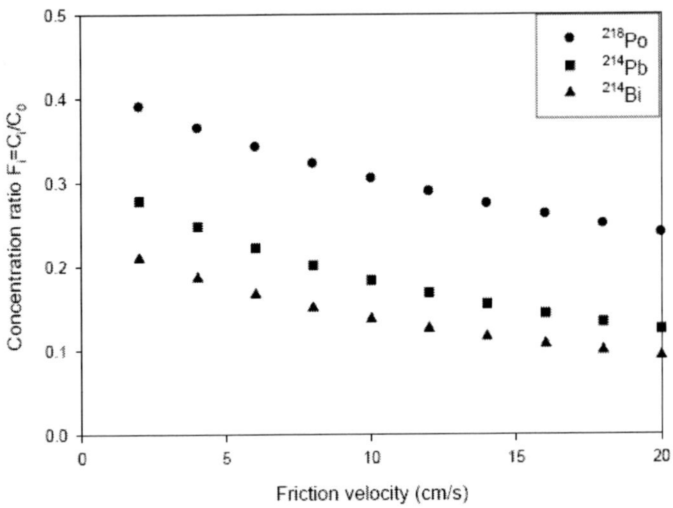

Figure 10. Concentration ratios of attached radon progeny in accumulation mode as a function on friction velocity.

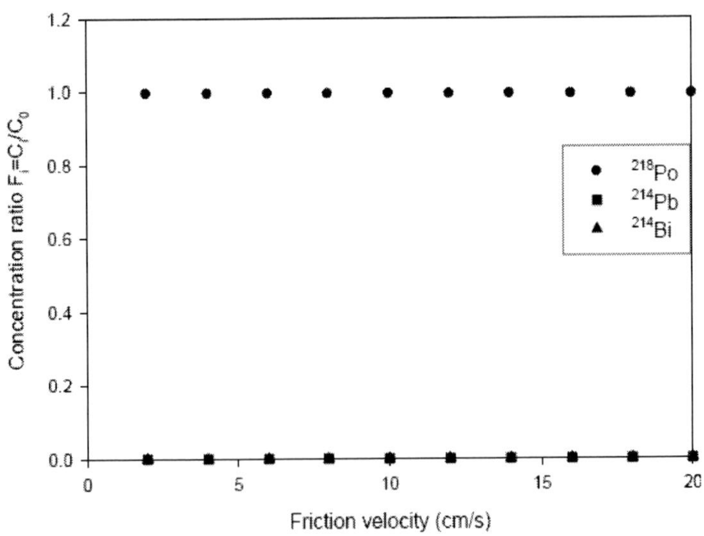

Figure 11. Concentration ratios of unattached thoron progeny as a function on friction velocity.

From the Figure 9 can be seen that unattached radon fraction decreases with friction velocity u*. It is because deposition rates increase with friction velocity and ventilation rate and larger ventilation causes larger airflow and faster air movement toward the room walls. Hence, unattached progeny atoms in air have smaller fractions with friction velocity.

The similar case is for attached radon fraction presented in Figure 10. It can be seen that concentrations of all progeny atoms strongly decrease with friction velocity.

It is very interest to see behavior of unattached thoron progeny with increasing of friction velocity in the room (Figure11). The ^{216}Po has a very short life time and it disappears on same position where is formed. From this reason, ^{216}Po atoms are in unattached form, almost 100% and their distribution is not changed due to ventilation and airflow.

References

Amgarou, K., Font, L., Baixeras, C., (2003). A novel approach for long-term determination of indoor 222Rn progeny equilibrium factor using nuclear track detectors. *Nucl. Instrum. Meth. Phys. Res.* A 506, 186–198.

Anderson R., (1988). Ventilation efficiency: analytical methods, scaling theory, and experimental techniques, in: Proceedings of the conference on Building systems: Room air and air contaminant distribution, ASHRAE, University of Illinois Press, Urbana-Cham-paign, IL, USA, pp. 195-199.

Choiniere, Y., Tanaka, H., Munroe, J.A., Suchorski-Tremblay, A. (1992), Prediction of wind induced ventilation for livestock housing, *Journal of Wind Engineering and Industrial Aerodynamics*, vol.41-44, pp.2563-2574.

Eisenbud M, Gesell T (1997) Environmental Radioactivity from Natural, Industrial, and Military Sources. San Diego, CA, Chapter 6 - Natural Radioactivity.

Fleischer R.L. (1997) Radon: Overview of properties, Origin and Transport. In: Durrani SA, Ilic R Radon measurements by etched track detectors. World Scientific, Singapore, 3, pp. 1-21.

ICRP 32 (1981) Limits for inhalation of radon daughters by workers, Pergamon Press, Oxford, Vol 6, Issue 1, 1-24.

ICRP 66, (1994). Human respiratory tract model for radiological protection. A report of a Task Group of the International Commission on Radiological Protection. ICRP publication 66. Ann. ICRP 24 (1–3) (Oxford: Pergamon Press).

Jacobi W (1972) Activity and potential *a* energy of ^{222}Rn and ^{220}Rn daughters in different air atmosphere. *Health Phys.* 22, 441-450.

Kato, S. (1997). *Appliance of CFD Technique for Designing Room Air Distribution — Part 1. Overview of CFD for Analyzing Indoor Climate.* Soc. Heating, Air-Conditioning and Sanitary Engineers, Japan, 71(6), 533–542.

Katase, A., Matshumoto, Y., Sakae, T., Ishibashi, K., (1988). Indoor concentrations of thoron and its decay products. *Health Phys.* 54, 283–286.

Lai, A.C.K., Nazaroff, W.W., (2000). Modeling indoor particle deposition from turbulent flow onto smooth surfaces. *J. Aerosol Sci.* 31 (4), 463-476.

Marsh, J.W., Birchall, A., (2000). Sensitivity analysis of the weighted equivalent lung dose per unit exposure from radon progeny. *Radiat. Prot. Dosim* 87, 167–178.

Marsh, J.W., Birchall, A., Butterweck, G., Dorrian, M.D., Huet, C., Ortega, X., Reineking, A., Tymen, G., Schuler, Ch., Vargas, A., Vezzu, G., Wendt, J., (2002). Uncertainty analysis of the weighted equivalent lung dose per unit exposure to radon progeny in the home. *Radiat. Prot. Dosim.* 102, 229–248.

Meckler G., (1992) HVAC system characteristics: ventilation and energy in: Proceedings of the conference on IAQ'92: Environments for people, ASHRAE, San Francisco, , pp. 327-338.

Mecker M. (1991), Indoor air quality design Guidebook, Fairmont Press, Liburn, GA, USA,.

Mercer IT. (1976)The effect of particle size on the escape of recoiling RaB atoms from particulate surfaces. *Health Phys.* 31:173-174.

Nikezic D. and Stevanovic N. (2007) Room model with three modal distributions ofattached 220Rn progeny and dose conversion factor, *Radiation Protection Dosimetry*, Vol. 123, No. 1, pp. 95–102

Nikolopoulos D, Vogiannis E. (2007) Modeling radon progeny concentration variations in thermal spas. *Sci. Total Environ.* 373:82–93;.

NCRP Report no. 93 (1987) Ionizing radiation exposure of the population of the United States. National Council on Radiation Protection and Measurements, Bethesda, Maryland.

Porstendorfer, J., Wicke, A., Schraub, A., (1978). A influence of exhalation, ventilation and deposition processes upon the concentrations of radon, thoron and their decay products in room air. *Health Phys.* 34, 465–473.

Porstendorfer J (1991) Tutorial lessons : Properties and behavior of radon and thoron and their decay products in the air. In. IV *Int. Symp. on the Nat. Radiat. Envir.*, Austria.

Yamasaki T, Guo Q, Iida T (1995) Distributions of thoron progeny concentrations in dwellings. *Radiat. Prot. Dosim.* 59, 135-140.

Porstendörfer, J., (1994). Properties and behaviour of radon and thoron and their decay products in the air. *J. Aerosol Sci.* 25 (2), 219-263.

Ramachandran TV, Sahoo BK (2009) Thoron (220Rn) in the indoor environment and work places. *Indian J. Phys.* 83 (8): 1079-1098.

Reineking, A., Knutson, E.A., George, A.C., Solomon, S.B., Kesten, J., Butterweck, G., Porstendorfer, J., (1994). Size distribution of unattached and aerosol-attached short-lived radon decay products: Some results of inter-comparison measurements. *Radiat. Prot. Dosim.* 56 (1–4), 113–118.

Steinhausler F (1996) *Environmental 220Rn*. A review. *Environ. Intern.*, 22 (1), S1111-S1123.

Stevanovic N, Nikezic D, Djordjevich A. (2004) The recoil factor of 214Pb. *J. Aerosol. Sci.* 35:1041–1050;.

Stevanovic, N., Markovic, V.M., Urosevic, V., Nikezic, D., (2009a). Determination of parameters of the Jacobi room model using the Brownian motion model. *Health Phys.* 96 (1), 48-54.

Stevanovic, N., Markovic, V.M., Nikezic, D., (2009b). Deposition rates of unattached and attached radon progeny in room with turbulent airflow and ventilation. *J. Environ. Radioact.* 100, 585-589.

Stevanovic N., Markovic V.M., Nikezic D., (2010) Relationship between deposition and attachment rates in Jacobi room model, *Journal of Environmental Radioactivity* 101 349-352

Straw, P.M., (2000). Computation and Measurement of Wind Induced Ventilation. Doctoral dissertation. University of Nottingham, UK.

Urosevic V, Nikezic D, Vulovic S (2008) A theoretical approach to indoor radon and thoron distribution. *Journal of Environmental Radioactivity* 99, 1829-1833.

UNESCAR Report (2000) Sources and effects of ionizing radiation. Volume I: Sources; Volume II: Effects, Unated Nations Scientific Committee on the Effect of Atomic Radiation, 2000 Report to General Assembly, with scientific Annexes. Unated Nations sales, Unated Nations, New York, Publication E.00.IX.3 and E.00.IX.4.

Zhao, B., Wu, J., (2006). Modeling particle deposition from fully developed turbulent flow in ventilation duct. *Atmos. Environ.* 40, 457-466.

Zhuo, W., Iida, T., Moriizumi, J., Aoyagi, T., Takahashi, I., (2001). Simulation of the concentrations and distributions of indoor radon and thoron. *Radiat. Prot. Dosim.* 93 (4), 3257–3368.

In: Ventilation: Types, Standards and Problems ISBN 978-1-61324-424-1
Editors: V. A. Romano and A. S. Duval, pp. 135-153

Chapter 5

ACOUSTIC FILTERS

Dehra Himanshu*

1-140 Avenue Windsor, Lachine, Québec H8R 1P7 Canada

ABSTRACT

The acoustic filter is used to filter unwanted frequencies of oscillations from a power system. It is a network with selective transmission for currents from a power system of varying frequency. The unwanted frequencies generated from a power system are removed by using an operational amplifier with different combination of filter arrangements. A new theory for noise protection and security from power systems is presented. A theoretical model for a phenomenon of photovoltaic amplification is formulated. The power systems are classified as per source signals of solar power, electric power, light power, sound power, heat power, fluid power and fire power. The acoustic filters as per source of noise signals are defined. The filters are differentiated as per source signal of unwanted frequencies from solar power, electric power, light power, sound power, heat power, fluid power and fire power. Some results from the simulation model and experiments of glass coated photovoltaic modules are presented. The full scale experiments for a glass coated photovoltaic duct wall section were conducted at Concordia University, Montréal, Québec. The full scale field performance of a photovoltaic duct wall is presented by characterisation of ventilation materials of climate, glass, solar panels, air, wood, polystyrene and electricity. Illustrated examples of noise filtering from power systems are: i) a vocal apparatus; ii) an airflow window with a photovoltaic solar wall; iii) a psychrometric air-conditioner; iv) a telephone line; v) a fire and smoke detection system; vi) sensors and transducers: energy balance for a human brain and vii) indoor environmental quality (IEQ): comfort and health.

Keywords: Source, ventilation materials, security, power system, signal, filter, sink, resonance.

---

*Corresponding Author. Tel: +1 514 366 3982, Email:anshu_dehra@hotmail.com.

1. INTRODUCTION

The displacement for any charged particle is defined by change of its position. A displacement has length and direction. The physical quantities that have features like displacements are called vectors. Vectors have both magnitude and direction and combine as per rules of addition. The physical quantities that are defined by number and unit and that only have magnitude are called scalars. The behaviour of gases over wide range of temperatures is predicted through kinetic theory model by average kinetic energy of translation.

In collisions with the source of an energy such as due to firing in air at the kinetic velocity of a bullet of a gun, the rotational and vibrational modes of motion are excited, which contribute to the internal energy of the air for propagation of sound waves. The total energy consists of kinetic energy of translation and kinetic energy of vibration of atoms in a molecule and potential energy of vibration of the atoms in a molecule. The magnetic energy also contributes to the total energy, however the available energy depends only on the temperature and has distribution of equal parts to each of the independent ways in which the molecules are able to absorb energy. The theorem (mention here, with no proof) is called equipartition of energy was deduced by James Clerk Maxwell. Each such independent way of absorbing energy is termed as degree of freedom for a molecule of gas.

The acoustic filters are defined based on the model of kinetic theory of gases for filtering unwanted frequencies of oscillations from a power system. It is a network with selective transmission for currents from a power system of varying frequency. A new theory for noise protection and security from power systems is presented. An acoustic filter is used to filter unwanted frequencies of oscillations from a power system. It is a network with selective transmission for currents from a power system of varying frequency. The noise protection and security is a crucial operation for obtaining a desired output from a power system.

A new theory for noise protection and security from a power system is presented. A description of filter systems for filtering noise is presented. The configurations of acoustic filters considered for illustrations are: i) the phonetics of filtering sound of human speech; ii) an airflow window fixed with a shading device and a photovoltaic (PV) solar wall for filtering daylight and solar power; iii) an air conditioner for filtering and conditioning of air and heat; iv) a telephone line for filtering an electrical noise; v) a fire and smoke detection system; vi) sensors and transducers: energy balance for a human brain; and vii) indoor environmental quality (IEQ): comfort and health.

2. THEORY

The unwanted frequencies generated from a power system are removed by using an operational amplifier with different combination of filter arrangements. The filters are differentiated as per source signal of unwanted frequencies from solar power, electric power, light power, sound power, heat power, fluid power and fire power. The acoustic filter is an electrical analog circuit of various combinations of RC feedback circuit with an operational amplifier [1, 2].

2.1. Operational Amplifier

An operational amplifier is an integrated circuit that consists of several bipolar transistors, resistors, diodes, and capacitors, interconnected so that amplification can be achieved over a wide range of frequencies. The open loop configuration of an operational amplifier has the highest possible gain when running wide open. In the closed loop configuration, it is easy to control the gain due to negative feedback. This feedback is obtained by connecting the output to the inverting input through a potentiometer. The negative feedback is useful for volume control of a highly versatile amplifier. The gain in closed loop configuration is directly proportional to the feedback loop resistance.

2.2. Acoustic Filters

The action of filtering the frequency from a power system is based on the variation in the reactance of an inductance or a capacitance with the frequency. The band of frequencies that can be removed from a power system can be at the low frequency end of frequency spectrum, at the high frequency end, at both ends, or in the middle of the spectrum. The filters to perform each of these operations are known respectively as low-pass filters, high-pass filters, band-pass filters and band-stop filters. There are many configurations of design of filters. The filters are divided into passive and active configurations. The passive filters are less effective simple circuits constructed with resistors, capacitors, and inductors. The active filters are useful in providing an effective filtering action than passive filters. The active filters require a source of operating power.

2.3. Acoustic Filter Systems

The criteria for definitions of filters for noise filtering is based on areas of energy stored in a wave due to noise interference, speed of wave and difference of power between two intensities of wave [4]. The filtered noise signals are considered from systems of solar power, electric power, light power, sound power, heat power, fluid power and fire power. The acoustic filters as per sources of noise are defined.

- Filter for noise of sol: The filter is used to filter noise due to difference of intensities of power between two solar systems. Example: window curtain, blind, wall and glass.
- Filter for noise of therm: The filter is used to filter noise due to difference of intensities of power between two heat power systems. Example: house, insulation, clothing and furnace.
- Filter for noise of photons: The filter is used to filter noise due to difference of intensities of power between two lighting systems. Example: electric bulb, television, computer and LCD screen laptop.
- Filter for noise of electrons: The filter is used to filter noise due to difference of intensities of power between two electrical power systems. Example: AM/FM radio

clock with ear phones, telephone instrument with ear phones and CD audio player with ear phones.

- Filter for noise of scattering: The filter is used to filter noise due to difference of intensities of power between two fluid power systems. Example: electric fan, pump, motor vehicle, river stream and tap water.
- Filter for noise of scattering and lightning: The filter is used to filter noise due to difference of intensities of power between two fire power systems. Example: cigarette lighter, gas stove, locomotive engine and thunder-bolt.
- Filter for noise of elasticity: The filter is used to filter noise due to difference of intensities of power between two sound power systems. Example: your vocal apparatus, organ pipe, thunder-bolt and drum beats.

2.4. Experiments and Simulation Model

The full scale experimental setup for a parallel plate photovoltaic device connected to a potentiometer was installed in an outdoor room facility located at Concordia University, Montréal, Québec, Concordia [3]. The schematic of the experimental setup is illustrated in Figure 1 [3]. An amplifier was built with a pair of glass coated photovoltaic (PV) modules forming a parallel plate duct with a plywood board and connected to a potentiometer. A potentiometer, a wire-wound variable resistor of up to 50 Ω was a wire-wound circular coil with a sliding knob contact [3]. It was used to vary electrical resistance across connected PV modules without interrupting the current. The characteristics of a parallel plate photovoltaic device connected to a potentiometer were established by varying electrical resistance with rotation of knob of a potentiometer. The current-voltage measurements were obtained for determining electric power output with a series electrical circuit connection of a pair of vertically inclined PV modules installed on a wooden frame.

The temperatures were measured as a function of volume of a parallel plate photovoltaic device. The non-linear thermal results include measurements of temperatures for PV modules, insulating panel and ventilated air column in the wooden frame. The air velocities were developed in the ventilated air column for transportation of heat both as a measure of buoyancy and fan induced hybrid ventilation. The electrical measurement results of currents, voltages and power with varying electrical resistance of potentiometer are presented in Table 1. The thermal measurement results of temperatures of various components of PV device, ambient air and room air temperatures, air velocities and solar intensities are presented in Table 2. The location and nomenclature of sensors is presented in Table 3. The results of the power output from a potentiometer with rotation of circular knob are illustrated in Figure 3.

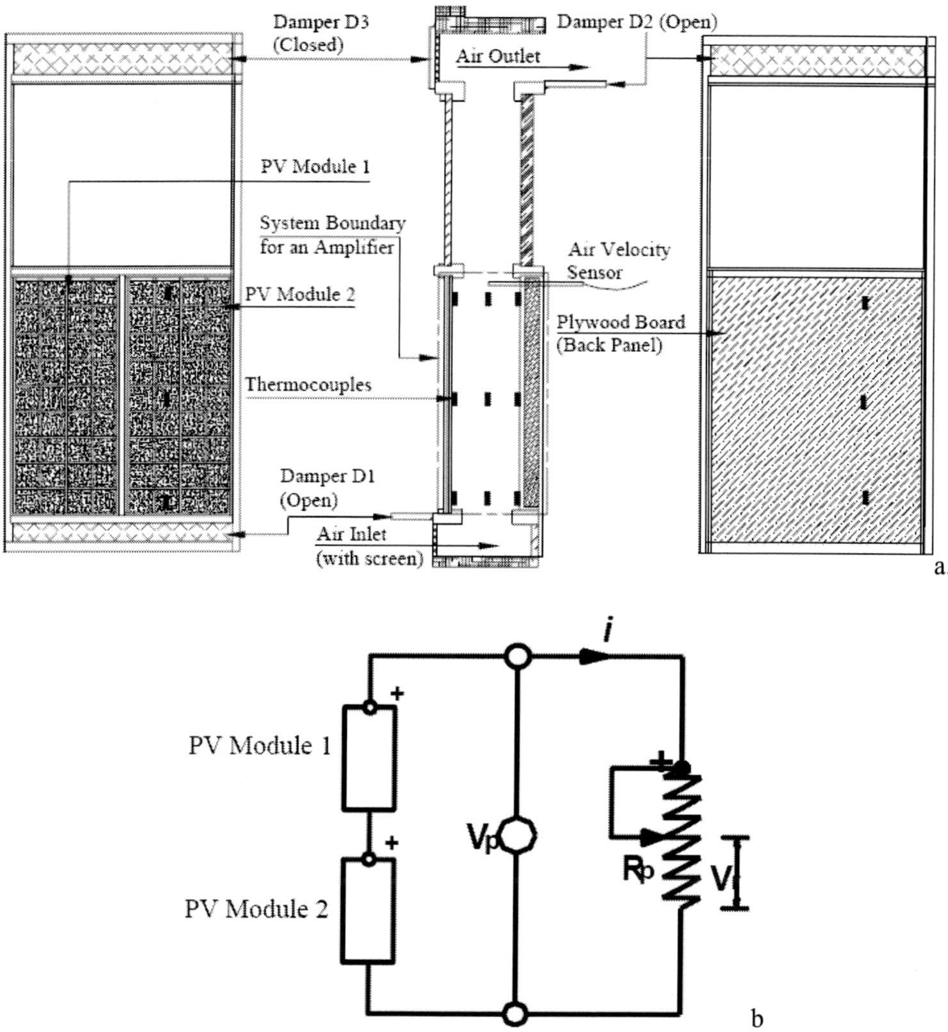

Figure 1. Schematic of experimental setup for a parallel plate photovoltaic device connected to a potentiometer: (a) location of sensors; (b) Electrical circuit diagram.

Table 1. Sample electrical measurement results with varying resistance of potentiometer

Rotation	Volts	Amps	Watts	Rotation	Volts	Amps	Watts	Rotation	Volts	Amps	Watts
240°	18.7	-	-	83°	16.3	0.935	15.23	30°	9.7	1.577	15.21
239°	16.5	0.331	5.461	75°	16.0	1.014	16.26	27°	9.0	1.587	14.33
201°	17.4	0.414	7.195	69°	15.8	1.100	17.38	21°	7.1	1.583	11.24
185°	17.5	0.454	7.940	64°	15.5	1.165	18.04	18°	6.2	1.573	9.831
162°	17.3	0.513	8.885	55°	15.0	1.302	19.53	17°	5.7	1.578	9.026
150°	17.18	0.550	9.449	50°	14.5	1.386	20.05	12°	3.9	1.567	6.257
142°	17.19	0.582	10.00	43°	13.2	1.503	19.79	10°	3.2	1.553	4.840
128°	17.1	0.640	10.93	42°	13.1	1.493	19.49	1.5°	0.5	1.593	0.807
107°	16.8	0.750	12.51	37°	11.9	1.536	18.26	1°	0.3	1.59	0.426
89°	16.4	0.884	14.45	32°	10.5	1.567	16.42	0°	-	1.643	-

Table 2. Sample thermal measurement results

Run No.	S (W m-2)	Ep (W)	To (° C)	Ts (° C)	v (m s-1)	Tp(b) (° C)	Tp(m) (° C)
1	697.5	31.0	13.5	22.1	0.451	34.5	33.01
2	725.4	31.1	15.9	22.9	0.362	32.5	33.3

Run No.	Tp(t) (° C)	Tb(b) (° C)	Tb(m) (° C)	Tb(t) (° C)	Ta(b) (° C)	Ta(m) (° C)	Ta(t) (° C)
1	36.2	20.2	24.4	27.6	18.7	19.3	22.5
2	35.7	20.2	23.9	29.1	18.3	19.7	23.3

Table 3. Nomenclature and location of sensors

Locations shown in Figure 1	$T_p(b)$ (° C)	$T_p(m)$ (° C)	$T_p(t)$ (° C)	$T_b(b)$ (° C)	$T_b(m)$ (° C)	$T_b(t)$ (° C)	$T_a(b)$ (° C)	$T_a(m)$ (° C)	$T_a(t)$ (° C)	Air velocity sensor
y (cm)	15	55	94	15	55	94	15	55	94	99
Z (cm)	60	60	60	60	60	60	60	60	60	60
X (mm)	6.2	6.2	6.2	96.2	96.2	96.2	51.2	51.2	51.2	51.2

Note: x is horizontal; y is vertical; z is adjacent 3rd axis of x-y plane.

A simulation model for prediction of temperature distributions varying with volume of a parallel plate photovoltaic device was developed [3]. The model has been used to predict temperature distributions at pre-defined locations in PV module, plywood board and air flowing through a parallel plate channel through the walls of vertically inclined photovoltaic modules. The results of the temperature plots are illustrated from Figure 3 (a) to Figure 3 (c).

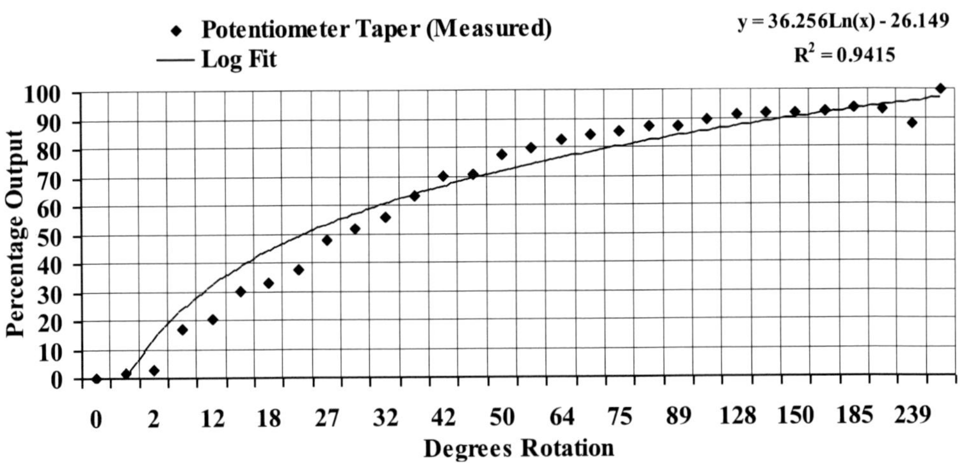

Figure 2. Potentiometer Taper (Measured) with percentage voltage output.

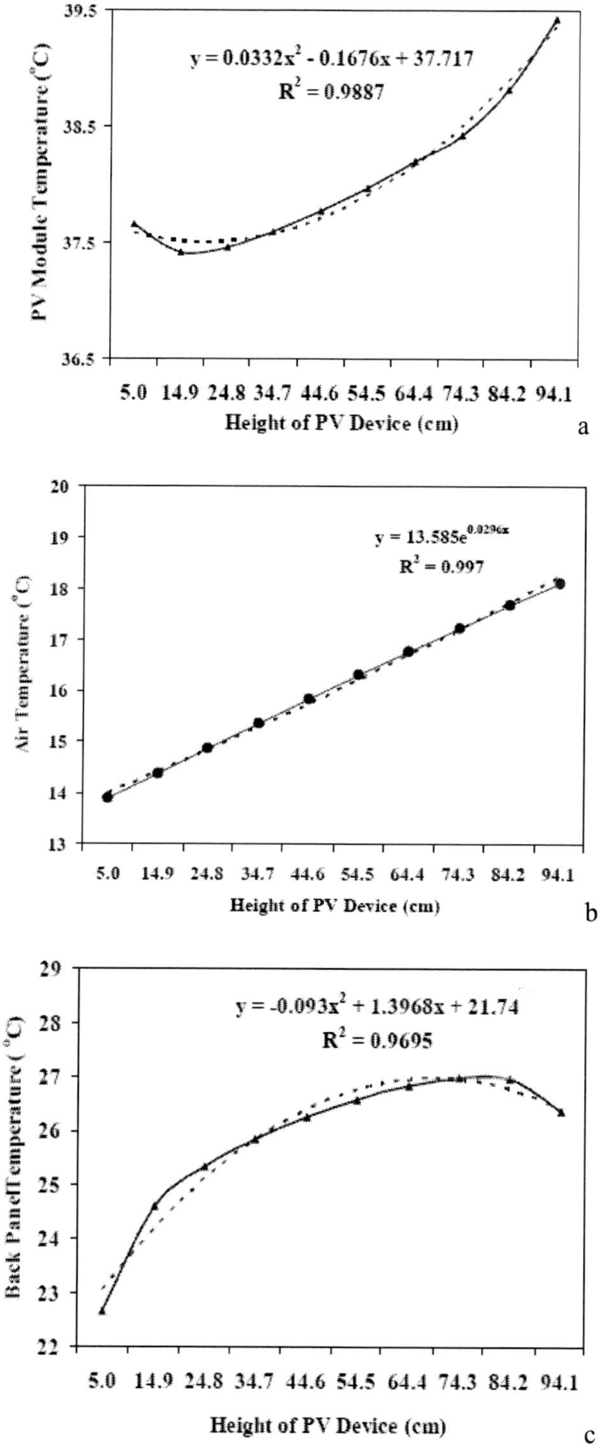

Figure 3. Temperature plots with height of PV device: (a) PV module; (b) air; and (c) plywood board.

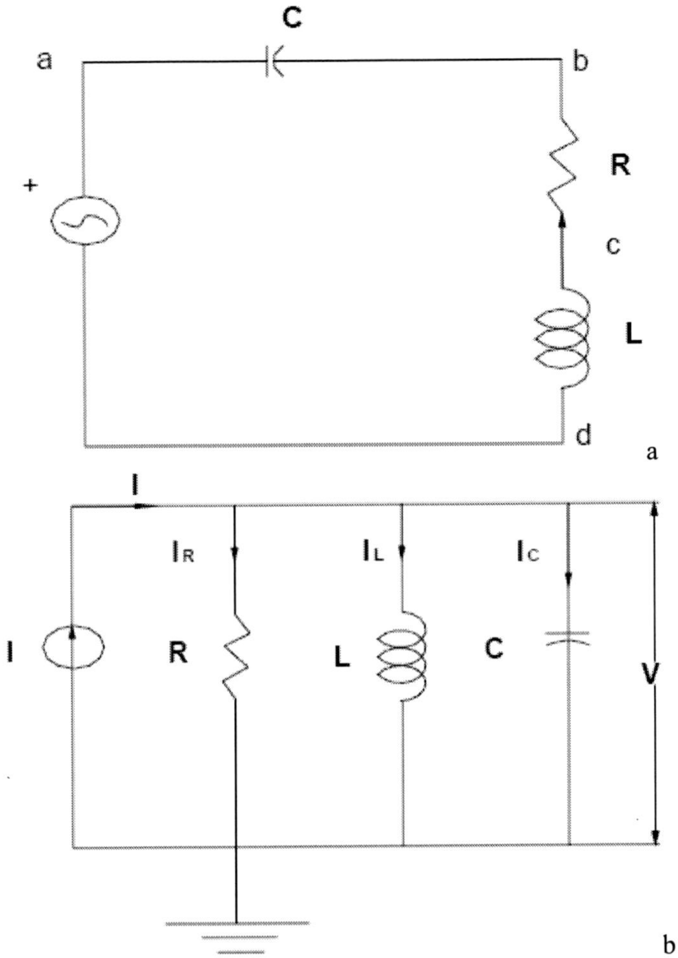

Figure 4. (a) L-C-R series arrangement of resonance; (b) L-C-R parallel arrangement of resonance.

3. Photovoltaic Amplification

The phenomenon of photovoltaic amplification has been observed from the graphs of Figures (2) and (3). The gain in steady state electrical and thermal functions for a photovoltaic device is a factor of its volume or resistance. This operational characteristic is similar to operation of a loudspeaker. The electrical analog has been used to describe the resonance phenomenon for equivalent mechanical, hydraulic and thermal systems of parallel plate photovoltaic device connected to a potentiometer. Figures 4 (a) and 4 (b) have presented series and parallel cases of L-C-R arrangement of resonance respectively.

3.1. Series Resonance

As is illustrated in Figure 4(a), for the case of series resonance, the equilibrium frequency w_o is written as:

$$w_o = \frac{1}{\sqrt{LC}} \tag{1}$$

The frequency w_o is called the resonant frequency. Its value is a function of two energy storing elements of the system. At resonance the impedance of the circuit is a minimum, and is equal to R. Consequently, when a series L-C-R circuit is at resonance, the current is a maximum and is also in time phase with the voltage. The power factor is unity. For operation at a frequency below w_o the resultant imaginary part of impedance is capacitive so that the current leads the voltage. When $w>w_o$, the inductive reactance prevails, so that the current then lags the voltage. When the ratio of inductive reactance to resistance in a series L-C-R circuit is high, a very rapid rise of current to the maximum value occurs in the vicinity of the resonant frequency.

3.2. Band Width

The mathematical procedure is used herewith for determining bandwidth at resonance. The bandwidth is defined as the range of frequency for which the power delivered to R is greater than or equal to $P_R/2$, where P_R is the power delivered to R at resonance. As the magnitude of the current at each half-power point during resonance is same, therefore:

$$i_1^2 R = \frac{1}{2} i_o^2 R = i_2^2 R \tag{2}$$

where, subscript 1 denotes the lower half-power point and subscript 2 the higher half-power point. The expressions for i_1 and i_2 are then written as:

$$i_1 = i_2 = \frac{i_o}{\sqrt{2}} = 0.707\, i_o \tag{3}$$

The bandwidth is defined during resonance as that range of frequency over which the magnitude of the current is equal to or greater than 0.707 of the current at resonance. For $w_oL/R=1/2$, the frequency at the lower half-power point is denoted by w1 and that of the upper half power point is denoted by w_2. Hence the bandwidth is w_2-w_1. Since the current at the half-power points is io/√2, it follows that the magnitude of the impedance must be equal to √2 R to yield this current. The reactance during resonance is at the lower half-power frequency X_1, is:

$$X_1 = w_1 L - \frac{1}{w_1 C} = -R \tag{4}$$

The minus sign appears on the right side of the equation because below resonance the capacitive reactance exceeds the inductive reactance. Rearranging Eq. (4), leads to:

$$w_1^2 + \frac{R}{L} w_1 - \frac{1}{LC} = 0 \tag{5}$$

The expressions for positive upper half-power and lower half-power frequency are:

$$w_1 = \frac{-R}{2L} + \sqrt{\left(\frac{R}{2L}\right)^2 + \frac{1}{LC}} \tag{6}$$

$$w_2 = \frac{R}{2L} + \sqrt{\left(\frac{R}{2L}\right)^2 + \frac{1}{LC}} \tag{7}$$

Therefore the expression for the bandwidth becomes, from Eqs. (6) and (7),

$$w_{bw} = w_2 - w_1 = \frac{R}{L} \tag{8}$$

3.3. Quality Factor Q_o of the RLC Circuit

By forming the ratio of the resonant frequency to the bandwidth, a quality factor to measure the selectivity or sharpness of tuning of the series RLC circuit is obtained [5]:

$$Q_o = \frac{w_o}{w_{bw}} \tag{9}$$

When Eq. (8) is inserted into Eq. (9):

$$Q_o = \frac{w_o L}{R} = \frac{X_{LO}}{R} \tag{10}$$

By making use of the knowledge that Q_o is very large in many resonant circuits and systems, the lower and upper half-power points are expressed in terms of the resonant frequency and the bandwidth.

The Eq. (6) is re-written as [5]:

$$w_1 = -\frac{R}{2L} + \sqrt{\left(\frac{R}{2L}\right)^2 + Q_o^2 \frac{R^2}{L^2}} \tag{11}$$

For values of Q_o which are 5 or greater very little error is made by writing [5]:

$$w_1 = -\frac{R}{2L} + \frac{R}{L} Q_o = -\frac{R}{2L} + w_o \tag{12}$$

Inserting Eq. (10) into Eq. (12) yields [5]:

$$w_1 = w_o - \frac{1}{2} w_{bw} \tag{13}$$

Similarly, expression for w_2 is written as [5]:

$$w_2 = w_o + \frac{1}{2} w_{bw} \tag{14}$$

The definition of quality factor is also expressed in terms of power storage. The instantaneous expression for the forced solution of current at resonance is:

$$i_o = I_{mo} \cos\left(w_o t\right) \tag{15}$$

The total power stored by the circuit in both L and C is:

$$W = \frac{1}{2} L i_o^2 + \frac{1}{2} C V_{CO}^2 \tag{16}$$

Inserting the appropriate expressions for i_o and V_{co} results in:

$$W = \frac{1}{2} L I_{mo}^2 \cos^{(2)}\left(w_o t\right) + \frac{1}{2} L I_{mo}^2 \sin^{(2)}\left(w_o t\right) \tag{17}$$

$$W - \frac{1}{2} L I_{mo}^2 = L I_o^2 \tag{18}$$

where, I_{mo} is maxima and Io is the rms value of current at resonance. The energy stored at resonance is a constant quantity. From Eq. (10), the expression for Q_o is written as [5]:

$$Q_o = \frac{w_o L I_o^2}{R I_o^2} = \frac{2 \pi f L I_o^2}{R I_o^2} = 2 \pi I_o^2 L \left(\frac{I_o^2 R}{f_o}\right)^{-1} \tag{19}$$

It is written in terms of stored energy and dissipated energy as:

Q_o=2πx(total energy stored/energy dissipated per cycle) (20)

It is applicable to any resonant system regardless of its composition.

3.4. Band Width for Parallel Resonance

For the parallel case, the equation for lower half power frequency w_l is written as:

$$w_1 C - \frac{1}{w_1 L} = -\frac{1}{R} \qquad (21)$$

The quadratic equation is written as:

$$w_1^2 + \frac{1}{RC} w_1 - \frac{1}{LC} = 0 \qquad (22)$$

Ignoring the negative solution, the desired solution is:

$$w_1 = -\frac{1}{2RC} + \sqrt{\left(\frac{1}{2RC}\right)^2 + \frac{1}{LC}} \qquad (23)$$

The expression for the upper half-power frequency is:

$$w_2 = \frac{1}{2RC} + \sqrt{\left(\frac{1}{2RC}\right)^2 + \frac{1}{LC}} \qquad (24)$$

The expression for the bandwidth is:

$$w_{bw} = w_2 - w_1 = \frac{1}{RC} \qquad (25)$$

The expression for quality factor is written as:

$$Q_o = \frac{w_o}{w_{bw}} = w_o RC \qquad (26)$$

4. Discussions on Theoretical Model

The incident short wavelength electromagnetic waves on a parallel plate photovoltaic device cause development of stresses and oscillations and results in propagation of following composite waves: i) transmission of electricity due to connected external load; ii) the heat waves are propagated at longer wavelength due to exchange of dissipated energy with medium of contact; iii) elastic waves are propagated in photovoltaic modules and polystyrene filled plywood board due to thermal stress development with propagation of heat waves; iv) fluid surface waves are propagated due to combined stress development with heat waves, elastic waves and applied external waves; v) applied external waves are propagated due fan induced pressure and climate-particle oscillations of wind.

Sound waves can propagate with combination of fluid surface waves and applied external waves such as in an organ pipe. The sound waves are also acting due to air-borne sound propagation from the surrounding environment. The propagation of acoustic waves is caused due to various stresses and oscillations of a static particle body. The stresses developed due to various forces acting on a cavity wall are classified as: i) fundamental stresses; ii) internal stresses; and iii) external stresses. The fundamental stresses are developed due to presence of gravitational and electromagnetic forces of a solar system. The internal stresses are developed under the influence of fundamental stresses and are defined by properties of a static particle body. The external stresses are developed under an external source of energy such as fan pressure used for active air ventilation through the cavity wall. The oscillations of a cavity wall are assumed to be generated because of stress development with cyclic load of day/night, heating/cooling and compression/expansion applied on a cavity wall. The climate-particle oscillations of wind are also assumed to be acting on a cavity wall. The acoustic waves of cavity wall are propagated due to stresses and oscillations of a cavity wall. The fluctuating forces are generated due to superposition of composite acoustic waves.

The analysis has been presented for series and parallel cases of L-C-R circuit resonance. The cases of resonance can be visualised with the aid of presented experimental and modelling results. Inductance is mass for the mechanical system for propagation of elastic waves. Capacitance is heat storage capacities of photovoltaic modules, air and polystyrene filled plywood board. Polystyrene filled plywood board is vulnerable to fire as soon as heat waves propagated with frequency matching with its latent heat of vaporisation is reached. Resistance is developed due to thermal and fluid resistance in energy storage elements of photovoltaic modules, air and polystyrene filled plywood board.

The series case of L-C-R resonance occurs with propagation of elastic waves of mechanical system of photovoltaic cavity wall. The parallel case of L-C-R resonance occurs with heat waves (C-R) and fluid surface waves (C-R) in conjunction with inductance (L) due to mass of photovoltaic cavity wall and resistance (R) due to temperatures of ground surface and ambient air.

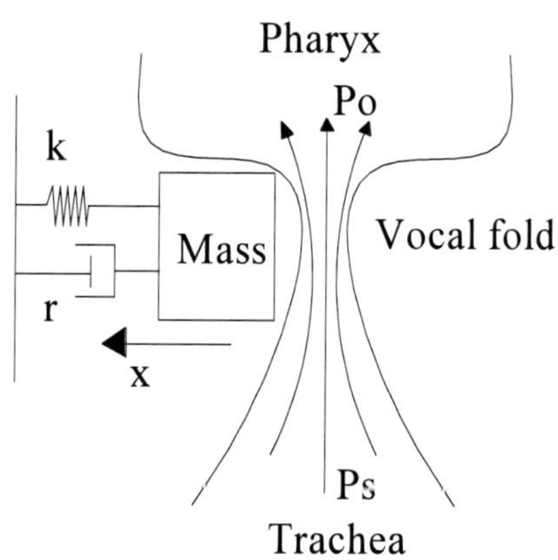

Figure 5. A vocal apparatus.

5. Advanced Level Configurations

5.1. A Vocal Apparatus

The example of phonetics of filtering sound of a human speech is illustrated. The human speech is synthesized due to development of stresses at vocal folds. The smoothening of the sound is function of its amplitude and its shape of oscillations at vocal tract of a human being. The vocal tract is a resonant cavity wall with sound energy stored in oscillations of its vocal folds [6, 7, 8]. The vocal apparatus showing mechanism of synthesis of human speech is illustrated in Figure 5.

5.2. An Airflow Window with a Photovoltaic Solar Wall

The filtering of solar energy is illustrated through an example of an airflow window attached with a shading device. An airflow window is fixed with a movable roller blind to control the transmission of daylight as well as amount of solar heat. The bottom portion of photovoltaic solar wall is used for controlling the amount of air ventilation along with generation of solar electric power. The vertically inclined photovoltaic duct wall was a built assembly of two glass coated solar panels and an insulating back panel of plywood board filled with polystyrene. The preliminary outdoor experiments were conducted for obtaining currents, voltages, temperatures, air velocities, sensible heat, thermal time constants and thermal storage capacity of glass coated solar panels installed on a wooden frame. The full scale experiments on characterisation of ventilation materials were conducted by taking into account the affect of materials of climate, glass, solar panels, air, wood, polystyrene and electricity. The example is illustrated in Figure 1 and Figure 6 [9, 10, 11].

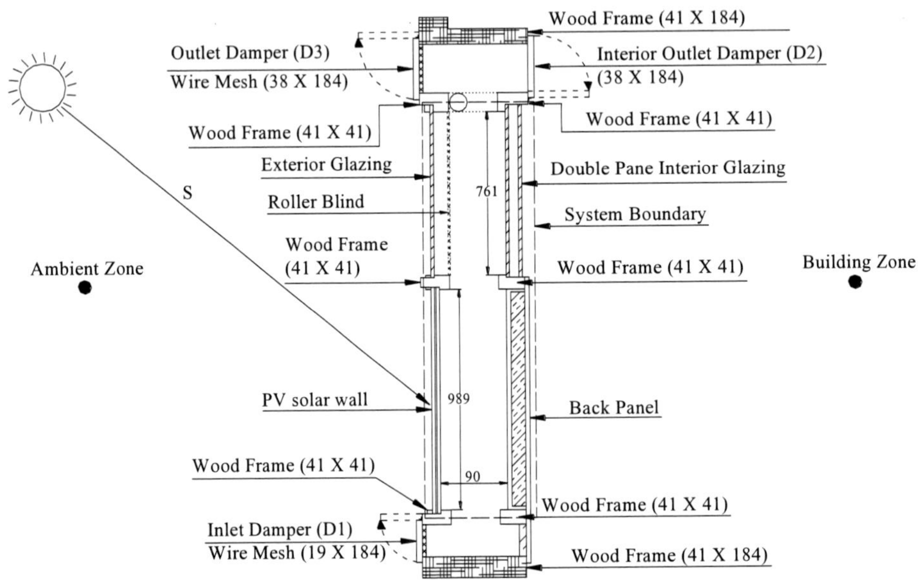

Dimensions shown are in mm.

Figure 6. An Airflow window.

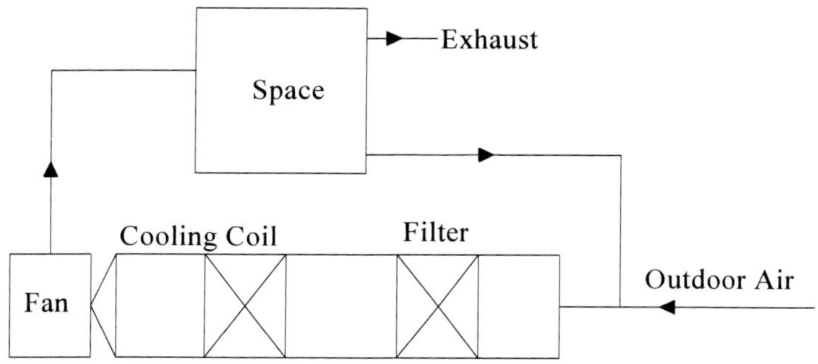

Figure 7. A psychrometric air conditioner.

5.3. A Psychrometric Air Conditioner

An elementary air conditioner for summer comfort conditioning consists of a cooling coil, a cooling fluid with a filter [12]. The schematic of operation of a psychrometric air conditioner is illustrated in Figure 7.

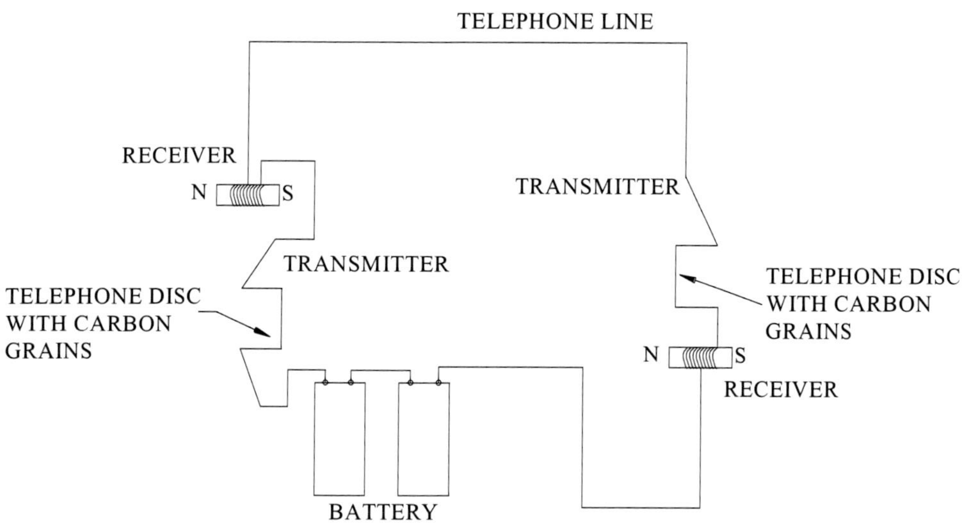

Figure 8. Operation of a Telephone Line.

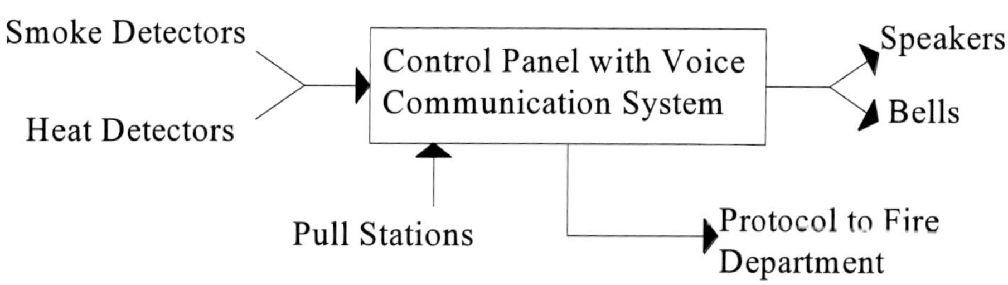

Figure 9. A fire detection system.

5.4. A Telephone Line

The impedance of a telephone line is composed of distributed resistance, capacitance, and inductance [13]. The impedance is a function of the length of the loop, the type of insulation of the wire, and whether the wire is aerial cable, buried cable, or bare parallel wires strung on telephone pole. A telephone line is usually supplied with a 48 VDC from the telephone exchange. The schematic of operation of a telephone line with telephone instrument is illustrated in Figure 8.

5.5. A Fire and Smoke Detection System

A fire detection system consists of a control system with interconnected alarms, smoke and heat detectors [14]. A fire detector is a device which is used for presetting an alarm at a particular temperature. A smoke detector is a device which is used for presetting an alarm when a certain percentage of smoke accumulates. The photovoltaic cell activates the smoke alarm only if it senses requisite obscuration of light over a unit area. The schematic of various components for fire detection system is illustrated in Figure 9.

5.6. Sensors and Transducers: Energy Balance for a Human Brain

Your body has feedback systems that regulate the internal environment of your body. The feedback systems make use of storage depots and numerous feedback loops. The monitoring of plasma calcium is a good example of negative feedback. The bones constitute large storage depots for calcium, for the plasma to withdraw these storage supplies in times of need. Our body's homeostatic regulatory systems are represented by feedback loops. The feedback is considered negative, when it is compensating or negates any change. The negative feedback is essential to stabilize a system.

The gastrointestinal tract, the lungs, the kidneys, and skin of your body make exchange of materials and energy between the internal and the external environments. A steady state is achieved by regulatory mechanisms involving the balance between the inflow and outflow of the internal environment that stabilizes the composition of the internal environment. The tendency to regulate the internal environment so that it is maintained in a steady state is called homeostasis.

The coordination of the activities of the various sensory organs within your body is controlled by the secretion of hormones and by transmission of nerve impulses. The nerve impulses travel along axons. The sensory nerve axons carry impulses from different parts of your body to the central nervous system (brain and spinal cord). Although parts of your brain involved in various physiological functions have been located, research attempts to localize complex behaviour patterns in sensory areas of your brain have failed. The sensory areas are located for the impulses on the cortex. The motor areas are also located where "command" impulses leave the cortex. The cerebellum receives impulses from sensory receptors and interacts with motor cortex, insuring smooth muscular movements. The reticular formation receives sensory impulses from the environment and sends impulses to the cortex.

The keeping of face beard (facial hair) and wearing of a knitted head cloth (*patka*) and a turban (*pag*) on your body has a logical and a scientific significance. The daily self-making folds of hair knots and making round folds of turban over the head of your body with colourful cotton cloths has following historical, medical benefits: i) it indicate, protects and concentrate the disciplinary physical and mental strength of a person; ii) it gives hair tonic to the growth of hairs on your body due to solar energy; iii) the whole system acts as an acoustic filter and provides immunity to your body; and iv) the folded *Patka* with style, folded design of hair knots on top of your head is your identity in time domain, the face beard on your body is a measuring ration and a sign of man, the turban with style, colour, design is your identity in space domain.

5.7. Indoor Environmental Quality (IEQ): Comfort and Health

The ASHRAE handbook of fundamentals [15] has provided a detailed discussion of the physiological principles of human thermal comfort. The amount of heat generated and dissipated by your body varies considerably with activity and age as well as with size and gender. The complex regulatory system of your body act to maintain the temperature of your body to about 36.9 °C regardless of the environmental conditions.

The environmental factors that affect thermal balance of your body for influencing thermal comfort are: i) the dry bulb temperature of the surrounding air; ii) the humidity of the surrounding air; iii) the relative velocity of the surrounding air; iv) the temperature of any surfaces that are in direct view to any part of your body and thus exchange radiation. In addition the persoal variables that affect the thermal comfort of your body are activity and clothing. The physiological mechanisms that your body uses to control your body temperatures are metabolism, blood circulation, near the surface of the skin termed as cutaneous blood circulation, respiration and sweating.

The metabolism rate determines the rate at which energy is converted from chemical to thermal form within your body and blood circulation controls the rate at which the thermal energy is carried forward to the surface of your skin. In respiration, air is taken in at ambient conditions but leaves saturated with moisture and very near the temperature of your body. The sweating has a significant effect on the rate at which energy is carried away from the skin of your body by heat and mass transfer.

The energy generated by metabolism rate of your body varies considerably with the activity of your body. A unit to express the metabolic rate per unit of area of your body is termed as met (1 met = 58.2 W m^{-2}), defined as the metabolic rate for your body while seated quite (called sedentary).

The variable which affects the comfort of your body is the type and amount of clothing that you are wearing. The insulation of clothing is defined as a single equivalent uniform layer over your whole body. The insulation value for clothing of your body is expressed in terms of clo units (1 clo = 0.155 m^2-C W^{-1}). A heavy business suit with accessories has insulation value of 1 clo, whereas a pair of shorts has 0.05 clo.

ASHRAE Standard 55 provides comfort conditions for an acceptable thermal environment. The most comfort studies involve use of the ASHRAE thermal sensdation scale. The scale relates words describing thermal sensations to a corresponding number. These are enumerated below:

+3	hot
+2	warm
+1	slightly warm
0	neutral
-1	slightly cool
-2	cool
-3	cold

The operative temperature is the average of the mean radiant and ambient air temperatures, weighted by their respective heat transfer coefficients. Energy balance equations are developed that use a predicted mean vote (PMV) index. The PMV index predicts the mean response of a large group of persons [16]. The coordinates of the comfort zones are [16]:

Winter. Operative temperature of 20 to 23.5 °C at 18 °C wet bulb temperature; operative temperature of 20.5 to 24.5 °C at 2 °C dew point.

Summer: Operative temperature of 22.5 to 26 °C at 20 °C wet-bulb temperature; operative temperature of 23.5 to 27 °C at 2 °C dew point.

The loss of hearing is a serious occupational hazard. Factory workers, truck drivers and other workers are exposed to noise levels that are harmful. The continous exposure to harmful noise level results in hearing imparement of your ear. The federal Occupational Safety and Health Administration (OSHA) in the Department of Labour has maintained noise exposure limits. The noise exposure is expressed in daily noise for an 8-hour shift.

CONCLUSION

A theory for noise protection and security with use of acoustic filters is presented. The acoustic filters for filtering noise from power systems are defined. The power systems are classified as per source signals of solar power, electric power, light power, sound power, heat power, fluid power and fire power. A phenomenon of photovoltaic amplification for a pair of vertically inclined glass coated photovoltaic modules is explained. The results of simulation model along with experimental results for noise filtering for a photovoltaic airflow window are presented. Some advanced level configurations of acoustic filters for different power systems are described.

REFERENCES

[1] Dehra, Himanshu. (2007). A unified theory for stresses and oscillations, *Canadian Acoustics*, Vol. 35, 132-133.

[2] Dehra, Himanshu. (2008). in Proceedings, American Institute of Chemical Engineers - AIChE 2008 Spring Meeting, Power transfer and inductance in a star connected 3-phase RC circuit amplifier, New Orleans, LA, USA, April 6-10, 2008, session 96a.

[3] Dehra, Himanshu. (2004). *A Numerical and Experimental Study for Generation of Electric and Thermal Power with Photovoltaic Modules Embedded in Building Façade*,

Ph.D. thesis, Department of Building, Civil and Environmental Engineering, Concordia University, Montréal, Québec, Canada, August 2004.
[4] Dehra, Himanshu. (2008b). The noise scales and their units, *Canadian Acoustics*, 36, 3, (September 2008) 78-79.
[5] Del Toro, V., (1986). *Electrical Engineering Fundamentals*, 2nd ed., Prentice-Hall of India, New Delhi, India.
[6] Randall, Robert H. (1951). *An Introduction to Acoustics*, Dover Publications, USA.
[7] Titze, Ingo R. (1994). *Principles of Voice Production*, Prentice Hall, Englewood Cliffs, NJ, USA.
[8] Lucero, J.C. (2005). Dynamics of the vocal fold oscillation, *TEMA Tend. Mat. Apl. Comput.*, 6, No. 1, 11-20.
[9] Dehra, Himanshu. (2007). in Proceedings, The 18th IASTED International Conference on Modelling and Simulation, On solar building energy devices, Montréal, Québec, Canada, May 30-June 1, 2007, 96-101.
[10] Dehra, Himanshu. (2008a). The entropy matrix generated exergy model for a photovoltaic heat exchanger under critical operating conditions, *International Journal of Exergy*, Vol. 5, Issue 2, 2008, 132-149.
[11] Dehra, Himanshu. (2009). A two dimensional thermal network model for a photovoltaic solar wall, Solar. Energy, Vol. 83, Issue 11, (November 2009) 1933-1942.
[12] J.L Threlkeld (1962). *Thermal Environmental Engineering*, Englewood Cliffs, NJ, USA.
[13] Dehra, Himanshu. (2007). in Proccedings, of 21st CANCAM, *A heat transmission model for a telephone line,* Department of Mechanical and Industrial Engineering, Ryerson University, Toronto, Ontario, Canada, June 3-7, 2007, 356-357.
[14] Newman, H. Michael (1994). *Direct Digital Control of Building Systems,* John Wiley and Sons, New York, USA.
[15] ASHRAE Handbook, *Fundamentals Volume* (1997), American Society for Heating, Refrigerating and Air-Conditioning Engineers, Inc., Atlanta, GA, USA.
[16] ANSI/ASHRAE Standard 55-1992 (1992). Thermal environmental conditions for human occupancy, American Society of Heating, Refrigerating and Air-Conditioning Engineers, Inc., Atlanta, GA, USA.

In: Ventilation: Types, Standards and Problems
Editors: V. A. Romano and A. S. Duval, pp. 155-172 ISBN 978-1-61324-424-1

Chapter 6

EVALUATION OF RANS AND LES TURBULENCE MODELS FOR INDOOR AIRFLOWS

Kiao Inthavong,[1*] Christian Heschl[2,†] and Jiyuan Tu[1‡]
[1]School of Aerospace, Mechanical and Manufacturing Engineering, RMIT University, Victoria, Australia
[2]Fachhochschulstudiengnge Burgenland, University of Applied Science, Pinkafeld, Austria

ABSTRACT

In buildings low indoor airflows persist, which presents difficulties in modelling of the airflow due to its transitional flow behaviour. The application of a number of turbulence models are investigated in order to evaluate their performance and to provide some issues related to indoor room modelling.

The most basic turbulence models are the RANS based models which are quick in providing a solution but are known to fail in transitional flows. Under different conditions such as a channel flow (to predict near wall turbulence), and free jets (to predict diffuser ventilation dispersion), the standard k-ε (SKE) model surprisingly provided sufficient results. Further experimental data was produced which allowed a calibration of the SKE model to further improve its performance. Because the SKE model is so widely used and is known for its generality, the calibrated model was further tested using an enclosed room, and was compared with a Large Eddy Simulation model and experimental data. Finally, the instantaneous results from the LES model highlighted the deficiency of steady simulations using RANS turbulent models which average and smear out the effects of local recirculating eddies.

* E-mail: kiao.inthavong@rmit.edu.au.
† Steinamangerstraße 21, A-7423 Pinkafeld, Austria. E-nail: christian.heschl@fh-pinkafeld.ac.at.
‡ Corresponding Author: Jiyuan Tu, PO Box 71, Plenty Road, Bundoora, Victoria 3083, Australia. Tel: +61-3-9925 6191, Fax: +61-3-9925 6108. E-mail: jiyuan.tu@rmit.edu.au.

INTRODUCTION

People spend most of their time in indoor environments such as residences, offices, schools, and workshops. In Australia, more than 90% of the workforce work in an indoor environment and more than 50% of the workforce is employed in office environments [1]. Therefore maintaining an acceptable level of indoor air quality and a thermally comfortable environment is of significant importance. This is achieved by ventilation systems that control indoor air parameters such as its distribution, velocity, temperature, and relative humidity. However a large portion of energy is consumed by buildings through heating, ventilation, and air conditioning (HVAC). A building's ventilation system is a major source of energy consumption and recently there is a push towards more efficient systems to reduce its contribution of greenhouse gases to the environment. In developed countries, it is estimated that 30~40% of energy is consumed in buildings and between 10% and 60% of this is used for air-conditioning and ventilation [2]. Buildings and building networks of the future will become energy self-sufficient, and the construction, operation, and deconstruction of buildings will need to meet the criteria of sustainability. To fulfill these requirements an accurate prediction of room airflows plays an important role in the design process of ventilation systems.

With advances in computational power, Computational Fluid Dynamics (CFD) is a promising tool for predicting indoor airflows. While its use has been prominent in recent years as reviewed by Spengler and Chen [3], Chen and Zhai [4], and Zhai [5], the accuracy and reliability of the simulation significantly depends on the CFD model settings. One of these settings is the use of turbulence models to account for the turbulent and transitional flow regimes that exist in indoor airflows. In this chapter some commonly used turbulence models found in many CFD software packages are evaluated for its turbulence kinetic energy in the near wall region, and its ability to model a jet spreading rate in an open room (typical of ventilation diffuser). The models are then applied to two room configurations and the results compared with experimental and LES data. All simulations are performed using the commercial CFD code Ansys-Fluent v12.1.

TURBULENCE MODEL DESCRIPTIONS

For engineering and industrial applications, airflow simulations will use the Reynolds Averaged Navier Stokes equations (RANS) with a first-order-closure turbulence models because of its lower computational effort. Generally most models use a linear correlation between the Reynolds stress tensor and the strain rate tensor known as the default Boussinesq approach. An eddy-viscosity (or turbulence viscosity) is then defined by the linear correlation and hence is also referred to as linear eddy-viscosity models. However this often fails to reproduce the turbulent normal stresses, particularly in areas of stagnation points [6] and in the prediction of turbulence driven secondary motions. Within room airflows both of these flow situations can be observed. Furthermore the air velocities found in ventilated rooms are often low so that the Reynolds number is in the order of $10^2 - 10^4$ which shifts in flow regimes from laminar to transitional and finally fully turbulent. In the transitional regime, entrainment of an inlet jet and its spreading rate are affected. Current RANS turbulence models are not able to consider such transitional effects without specific modifications to the turbulence models. To close the RANS equation the Reynolds stresses $-\rho\overline{u_i^{'}u_j^{'}}$ (which represent the turbulence effects) are modelled by including additional transport equations. These additional model equations make up the turbulence model.

k-ε Models

The family of *k*-ε models first proposed by Launder and Spalding [7] are known for its generality in producing reasonable accuracy for a wide range of turbulent flows. It uses two additional transport equations,

$$\rho u_i \frac{\partial k}{\partial x_i} = \frac{\partial}{\partial x_j}\left[\left(\mu + \frac{\mu_t}{\sigma_k}\right)\frac{\partial k}{\partial x_j}\right] + P_k - \rho\varepsilon \tag{1}$$

$$\rho u_i \frac{\partial \varepsilon}{\partial x_i} = \frac{\partial}{\partial x_j}\left[\left(\mu + \frac{\mu_t}{\sigma_\varepsilon}\right)\frac{\partial \varepsilon}{\partial x_j}\right] + \frac{C_{\varepsilon 1}P_k - C_{\varepsilon 2}\rho\varepsilon}{T} \tag{2}$$

in order to determine the turbulent velocity and time scales as:

$$T = \frac{k}{\varepsilon} \qquad v = \sqrt{k}$$

and the turbulent viscosity is expressed as

$$\mu_t = \rho c_\mu \frac{k^2}{\varepsilon}$$

k is the turbulence kinetic energy, ε is the dissipation rate of turbulence energy, and C_μ = 0.09 is the default empirical constant. The standard *k*-ε model is a semi-empirical model with its derivation based on high Reynolds number flows primarily valid for flow regions far from wall boundaries (turbulent core flow). Thus for low Reynolds number flows, such as those found near walls, wall functions [7] are typically used to connect the turbulent core flow with and the near-wall flow. Modifications to improve the standard k-ε (SKE) model include the:

- Realiable k-ε model (RKE) by Shih et al. [8] which aims to improve predictions of the spreading rate of both planar and round jets and improved accuracy for flows involving rotation, boundary layers under strong adverse pressure gradients, separation, and recirculation.
- Re-Normalization Group k-ε model (RNG-ke) by Yakhot and Orszag [9] which aims to improve the accuracy for rapidly strained flows and high swirling flows. In addition the turbulent Prandtl numbers are analytical derived instead of user-

specified, constant values. Furthermore the RNG theory accounts for low-Reynolds-number effects provided that the mesh in the near wall region is sufficiently fine.

k-ω Models

The family of *k-ω* models have also become more popular in many industrial applications given the large improvements made to the model by Wilcox [10] and Menter [11]. Compared to the k-ε model, the k-ω model can better predict adverse pressure flows [10], but does not predict free shear flows well. The k-ω Shear Stress Transport (SST) model [12] provides a transition from the inner region of the boundary layer using k-ω equations, to the turbulent core region that uses transformed k-ε model equations. The transition is performed by the model's own blending functions and hence there is no wall treatment for the k-ω models and instead it requires a fine mesh to resolve the near wall regions.

v^2-*f* Models

The Boussinesq assumption in the standard k-ε turbulence model is applied together with wall functions. But it is well known that in the near wall region the standard *k-ε* turbulence model must be modified via damping or wall functions. These functions are usually derived for simple flows, such as channel flows, and are ill-suited for most complex flows. Based on this observation Durbin [13] introduced an alternative eddy viscosity formulation, v^2-*f* -turbulence model, which uses an imaginary turbulent normal stress component v'^2. This imaginary stress component is always normal to the closest wall and represents the turbulent velocity scale. Two further transport equations are introduced on top of Eqn (1) and (2). These are:

$$\rho u_i \frac{\partial v^2}{\partial x_i} = \frac{\partial}{\partial x_j}\left[\left(\mu + \frac{\mu_t}{\sigma_\varepsilon}\right)\frac{\partial v^2}{\partial x_j}\right] + \rho k f - 6\frac{v^2}{k}\rho\varepsilon \tag{3}$$

$$L^2 \frac{\partial^2 f}{\partial x^2_j} - f = \frac{C_1}{T}\left(\frac{v^2}{k} - \frac{2}{3}\right) - C_2\frac{P_k}{k\rho} - \frac{1}{T}\left(6\frac{v^2}{k} - \frac{2}{3}\right) \tag{4}$$

In the v^2-*f* model the scales of turbulence is determined as:

$$T = \max\left(\frac{k}{\varepsilon}, 6\sqrt{\frac{v}{\varepsilon}}\right), \qquad L = C_L \max\left(\frac{k^{3/2}}{\varepsilon}, C_\eta \frac{v^{3/4}}{\varepsilon^{1/4}}\right) \tag{5}$$

and the turbulent viscosity is expressed as

$$\mu_t = \rho c_{\mu,v2f}\overline{v'^2}T \tag{6}$$

The transport equation for the normal-to-wall fluctuation velocity can be derived from the Reynolds stress equation with linear pressure strain model and the boundary layer simplifications. In regions far away from the wall v'^2 gets too large such that $\overline{v'^2} > \frac{2}{3}k$ which is physically incorrect. For this reasons Davidson et al. [14] proposed a simple modification to set an upper bound for the source term kf in the v'^2 equation. The modification ensures that $v'^2 < 2/3k$. In regions where $v'^2 \approx 2/3k$ the v'^2 model predicts the turbulent viscosity considerably larger as the standard k −ε turbulence model so that Davidson et al. [14] suggested a simple remedy to compute the turbulent viscosity:

$$\mu_t = \min\left\{\rho C_{\mu,k\varepsilon}\frac{k^2}{\varepsilon}, \rho c_{\mu,v2f}\overline{v'^2}T\right\} \tag{7}$$

This model will be referred as V2F-DAV hereafter.

RSM Models

The Reynolds Stress Model (RSM) moves away from the eddy viscosity assumption which produces isotropic turbulence structures. This leads to poor accuracy for flows with strong anisotropic properties, such as swirling flows and flows with strong curvatures. Transport equations for each of the Reynolds stresses, together with an equation for the dissipation rate are introduced. This means that there are five additional transport equations in 2D flows, and seven in 3D. This leads to more intensive computational effort in order to capture better more complex flows. For design purposes of indoor airflows, the increase in fidelity may not be a good compromise for the longer time taken per simulation. In general the pressure strain term in the RSM can be computed by the following decomposition:

$$\phi_{ij} = \phi_{ij,1} + \phi_{ij,2} + \phi_{ij,w} \tag{8}$$

The first term $\phi_{ij,1}$ refers to the "turbulence interaction" part of the pressure strain process and is called "the slow pressure strain term". This term describes the interaction between the pressure and velocity fluctuations. The second term $\phi_{ij,2}$ is called the mean strain or "rapid" contribution for the pressure strain process and describes the interaction between the pressure and the time-averaged velocity field. If the flow regions are influenced by rigid walls the third term $\phi_{ij,w}$ must be included. This term considers the reflection of pressure fluctuations near walls which are responsible for the redistribution of normal stresses.

In the commercial CFD code Ansys-FLUENT two pressure-strain RSM models are available. The most commonly used model is the linear pressure strain model (IP) according to the proposal by Gibson and Launder [15], Fu et al. [16] and Launder [17]. In near wall regions a low Reynolds modification by Launder and Shima [18] is also available.

Evaluation of Turbulence Models for Indoor Air Flows

Table 1 summarises the turbulence models chosen for its performance in indoor air flows.

The turbulence structures at the near wall region are obtained by simulating a turbulent channel flow with a $Re_\tau = 590$ and compared with the DNS of Moser [19]. Because some models use wall functions and some use a low Reynolds enhanced treatment two different grids were designed. The grid for the low-Re models was designed with a $y+ < 1$ and the other grid with a $y+ \approx 60$.

Table 1. Summary of turbulence models for evaluation

Turbulence model	Shortname	Comments
standard k-ε model	SKE	standard model in Fluent
RNG k-ε model	RNG-KE	standard model in Fluent
realizable k-ε model	RKE	standard model in Fluent
k-ω model	KW	standard model in Fluent
k-ω model with shear flow correction	KW-SFC	standard model in Fluent
RSM with linear pressure strain model and wall reflection term	RSM-IP	standard model in Fluent
v^2-f modified by Davidson et al. [14]	V2F-DAV	User Defined Function model

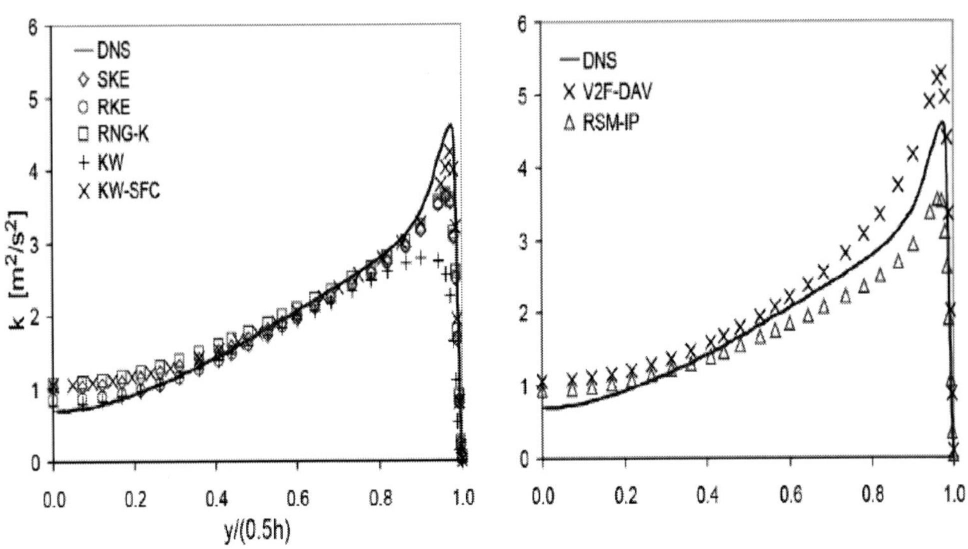

Figure 1. Turbulent kinetic energy (TKE) profile in a turbulent channel flow at Re = 590. The value y/(0.5h)= 0 on the abscissa denotes the channel centre.

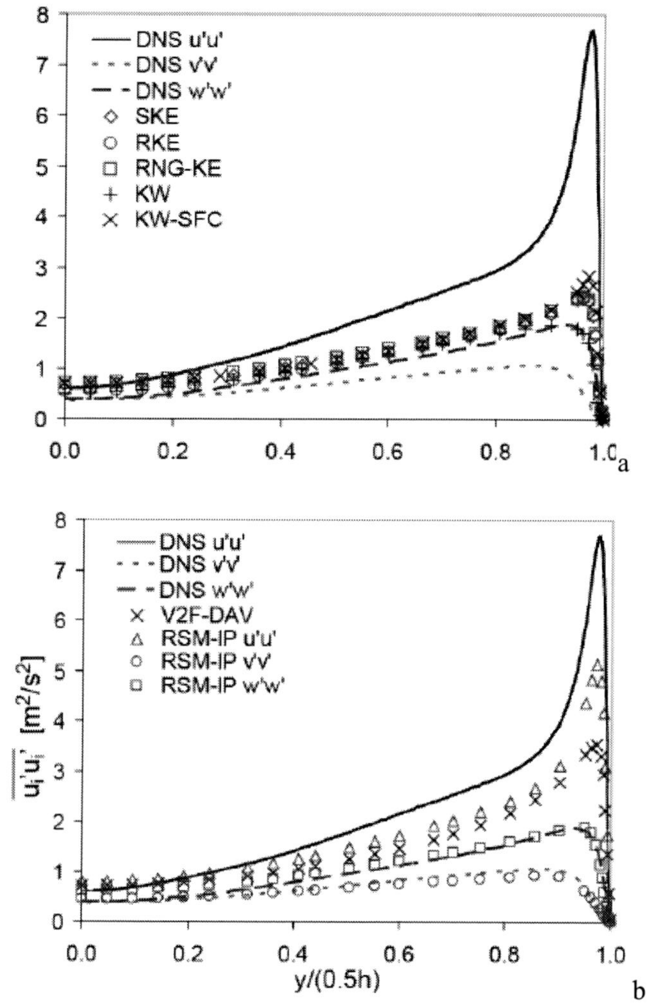

Figure 2. Reynolds stress profiles in a turbulent channel flow at Re = 590. The value y /(0.5h) = 0 on the abscissa denotes the channel centre.

Two Dimensional Turbulent Channel Flow

Figure 1 shows the computed results for the turbulent kinetic energy (TKE) profile in the near wall region of a 2D turbulent channel flow. The k-ε models reproduce a similar TKE profile closely matching the DNS data until very close to the wall. In the near wall region, it under predicts the TKE. The k-ω also reproduces a similar profile in the turbulent core region but under predicts the TKE even more than the k-ε model in the near wall region. The k-ω SST model with its blending function of improves the profile in the near wall and overall has the best matching results. The near wall behaviour is generally predicted well by both the v2f-dav and RSM-IP models. The v2f model over predicts the peak in the turbulent kinetic energy while the RSM model under predicts this.

The Reynolds stresses $\overline{u_i'u_i'}$ are anisotropic which means that only the RSM model will directly output each Reynolds stress. For the RANS models that are based on a linear eddy-viscosity assumption, decomposition of the turbulent kinetic energy into its fluctuating velocity components by $u' = v' = w' = (2k/3)^{0.5}$ can be applied. This implies isotropic turbulence and therefore the Reynolds stress profiles collapse to one single curve as seen in Figure 2a. The profiles for the *k-ε* and *k-ω* models show a similar profile to the *w'w'* Reynolds stress, which are significantly less than the streamwise Reynolds stresses *u'u'* but also significantly greater than the Reynolds stresses in the wall normal direction, *v'v'*. Similarly the v2-f model exhibits a similar trend however its peak at the near wall is larger than the k-ε and k-ω models. The RSM model shows good agreement for *v'v'* and *w'w'* but under predicts the peak and the general profile of the *w'w'* Reynolds stress.

Jet Spreading Rate

In ventilated rooms usually low inlet velocities are used to avoid draught in the occupied zone. Therefore low Reynolds jets in the Reynolds number range of 10^2-10^4 are often used in the inlet area. This has a far-reaching consequence since most turbulence models have difficulty in reproducing such transitional effects. Therefore simulations of a two-dimensional wall jet are performed and the spreading rate produced by the different turbulence models is compared with the experimental data from Deo [20]. In this set up a two-dimensional wall jet is fed through a slot immediately adjacent to a wall (Figure 3). The flow consists of an inner layer which is between the wall and the centre of the stream velocity akin to a boundary layer flow, and an adjacent outer layer, which corresponds to a free shear flow similar to the two-dimensional turbulent free jet.

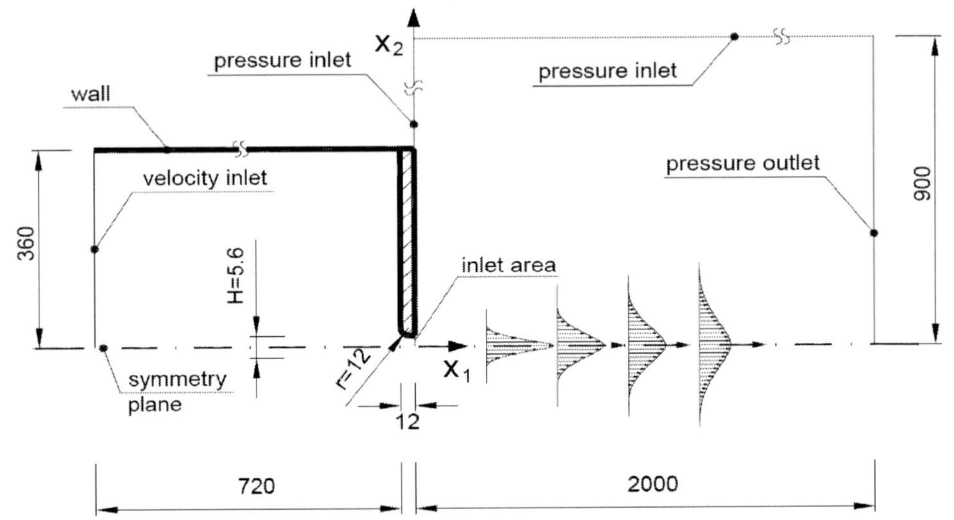

Figure 3. Measurement setup of Deo [20] and derived computational domain (dimensions in mm).

The spreading rate of different turbulence models over a range of Reynolds number is shown in Figure 4. The measured spreading rate shows a decreasing trend as the Reynolds number increases. In contrast, the spreading rate for the different turbulence models used show no variations for the increase in Reynolds number.

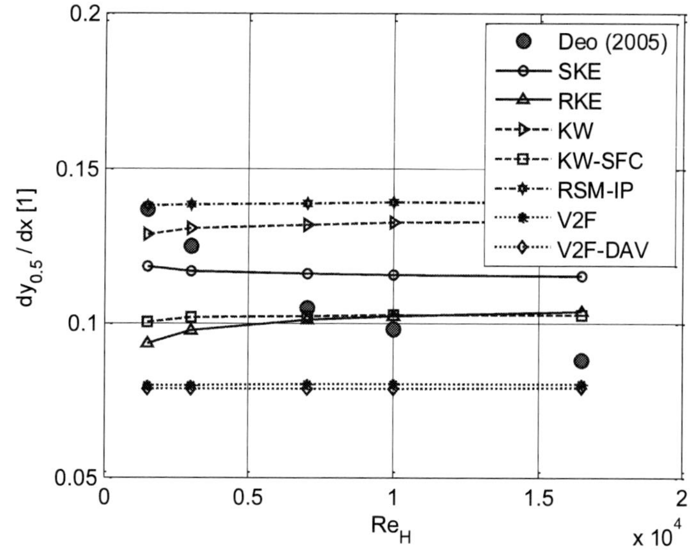

Figure 4. Spreading rates computed over a range of Reynolds numbers for different turbulence models.

The range of the different computed spreading rates is almost within the band of the measured data. Here the best choice for the low Reynolds number jets appear to be the RSM-IP, KW and the SKE models and for high Reynolds number jets the V2F, KW-SFC and RKE models. Therefore the continuation of this chapter will focus on the improvements of the best performing models at the low Reynolds number jets. Currently, RANS turbulence model do not handle transitional behaviour of free jets very well, and therefore individual adjustment of the model constants for the RSM-IP, KW and the SKE is necessary to reproduce the different spreading rates.

Calibration of Turbulence Models

To improve the turbulence modelling of indoor airflows, specifically flows involving jet streams produced by ventilation ducts and/or diffusers, individual tuning of model constants can be performed with experimental data for a case under investigation. It is known that the tuned model constants must be interdependent and consistent to the log-law layer [10, 21]. Therefore a simple option is to adapt the proportionality factor of the production and dissipation term in the ε or ω equation in such a way that the spreading rate and the logarithmic wall law are fulfilled. According to k-ε and k- ω based turbulence models the following link between the model parameters should be ensured:

$$k^2 = (C_{\varepsilon 2} - C_{\varepsilon 1})\sigma_\varepsilon \sqrt{C_\mu} \tag{9}$$

$$\alpha_\infty = \frac{\beta}{\beta_\infty^*} - \frac{k^2}{\sigma_\omega \sqrt{\beta_\infty^*}} \tag{10}$$

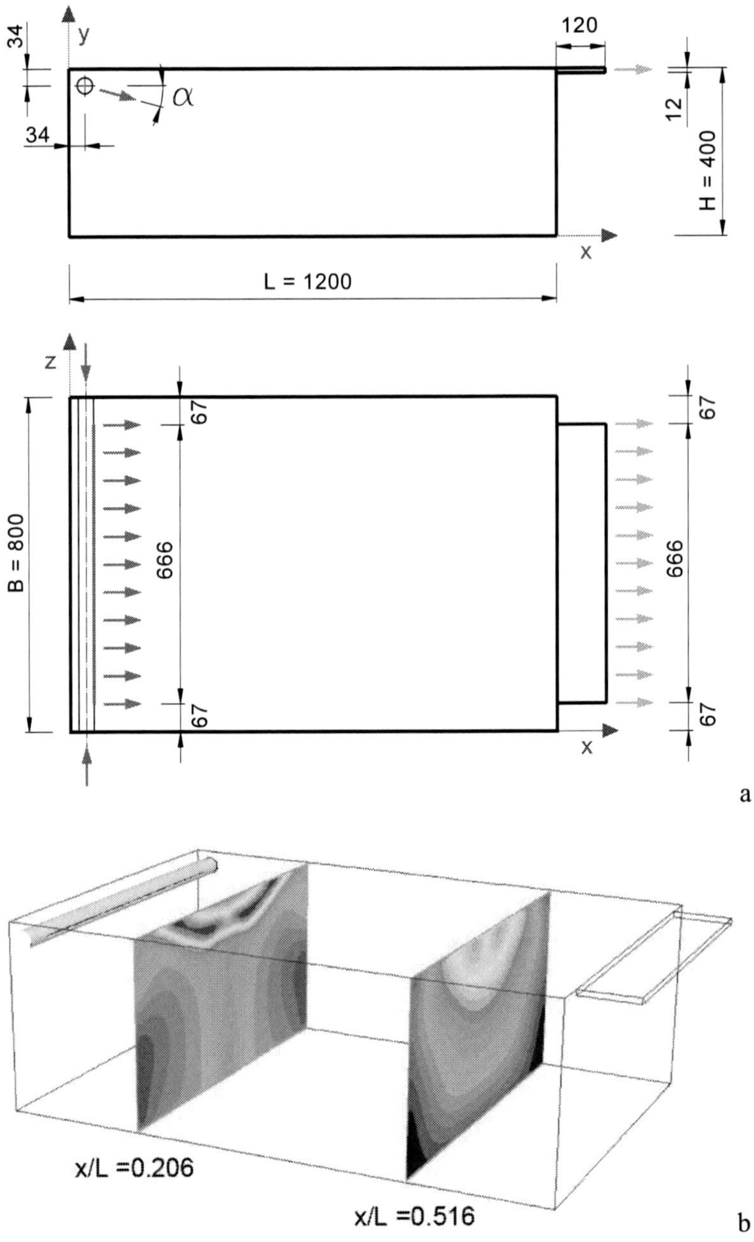

Figure 5. a) Geometry and dimensions of the plexiglass model room for Case 1. b) Planes x/L =0.206 and x/L=0.516 in relation to the model room used for post processing.

If the spreading rate is known, a specific calibration of the model constants based on the equation (9) and (10) is feasible. Thus room airflow experiments are performed to measure the mean inlet velocity and its spreading rate. The scaled room model and its dimensions are shown in Figure 5. The air supply duct (38 x 3mm) has evenly distributed holes and is fed from both side ends over a flow rectifier. The air supply duct has 7 groups of 32 holes with a diameter of 1.3 mm and an overall length of 440.8 mm. The distance between the holes is 1.5 mm and between the seven-hole groups 13.5 mm. The mean inlet velocity is 36.88 m/s and the inlet angle is 27.5° measured to the horizontal direction. The experimental spreading rate for the room was found to be $dy_{1/2}/dx = 0.108$. The model constants are tuned, based on Equation (9) and (10) to yield $C_{\varepsilon 1}$=1.42, σ_ε=1.23 and is subsequently referred to as SKE-NL (standard *k-ε* – non linear). The other model constants conformed to the standard default values.

Case 1: Application of Turbulence Model with Experimental Data for a 2D Room Airflow

The turbulence models including the tuned SKE-NL model is applied to the same 2D room airflow used for the experiment in Figure 5. For the CFD model two grids were used; one with 2.5million cells with a $y^+ \sim 30$ for turbulence models using a standard wall function, (SKE, SKE-NL, RKE, and RSM-IP models) and the other with 3.5million cells with a $y^+ < 1$ (KW-SFC). The measured and predicted CFD results for the velocity distribution is shown in Figure 6. It can be seen that the flow contains a free jet at the inlet region (plane at x/L = 0.25) and downstream a wall affected region (plane at 0.75). In the wall affected area the flow is influenced by the Coanda effect which forces the jet to remain close to the ceiling. The measured results show that the flow is nearly symmetrical which is captured by most of the used turbulence models.

Only the KW-SFC and partly the RKE model show an asymmetrical flow pattern. This is due to the turbulence interaction which is too dissipative so that the entrainment is significantly underpredicted (the measured spreading rate in the inlet area is about $dy_{1/2}/dx = 0.11$ and the predicted $dy_{1/2}/dx = 0.09$. This leads to a lower entrainment in the shear layer flow and consequently to a smaller jet profile with a higher centreline velocity. This behaviour of the RKE and KW-SFC model can be seen where a complete attachment of the jet to the wall is suppressed. Surprisingly the simple SKE model with a computed spreading rate of $dy_{1/2}/dx = 0.11$ predicts the flow pattern and the velocity distribution better than the RKE, KW-SFC and RSM-IP model. The computed flow pattern indicates that an accurate prediction of the entrainment is very important. Because of the low inlet velocity a transitional flow behaviour, which leads to a higher entrainment effect (Figure 4), can be expected.

Accordingly, the model constants of RANS-based turbulence models should be adjusted before they are used for room airflow simulations. The impact of the nonlinear correlation between the anisotropy and the strain rate and rotation rate tensor can be seen from the velocity distribution in Figure 6. Comparison of the velocity distribution between the SKE and the SKE-NL shows that the nonlinear model provides improved results with the experimental findings.

This is most evident for regions far away from the inlet (x/L=0.75) where the flow pattern of the wall jet at the ceiling is more realistically reproduced. The improved normal stress distribution of the nonlinear model (see Figure 9) enables turbulence driven secondary motion, which improves the prediction of the lateral spreading rate of the three dimensional wall jet on the ceiling. In comparison with the differential Reynolds stress model (RSM-IP) the nonlinear eddy viscosity model gives comparable results but needs less computational effort.

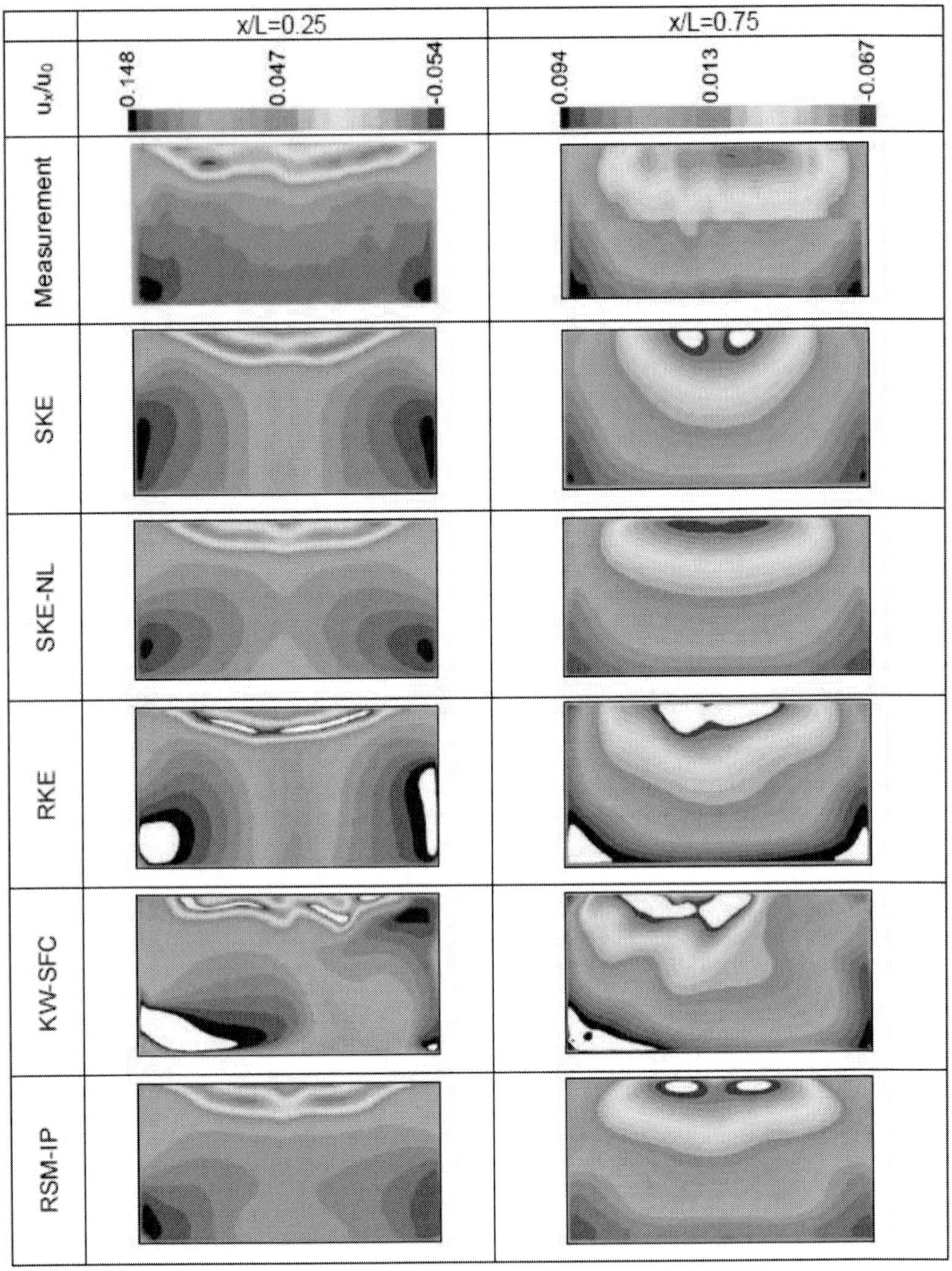

Figure 6. Comparison of measured and computed velocity distribution in model.

Case 2: Application of Turbulence Model with Experimental and LES Data for Room 2

In this section we introduce the more sophisticated turbulence model, Large Eddy Simulations (LES). This approach is much more computationally intensive than RANS turbulence models because the large eddies are solved directly and the small eddies are modelled by a subgridscale (SGS) model. As a result, LES is less influenced by modelling errors than the RANS models. Furthermore, LES accounts for the history and transport effects on turbulence in which RANS model is incapable of due to its inherent time-averaging modeling approach.

One concern of LES is the large computational expense that has prevented the use of the LES in the past. However, this section shows that LES computations can be carried out on current computer workstations within reasonable computational times. It is therefore expected that, with the progress of computer speeds, the time expense may be tremendously reduced and LES may become a popular tool to study indoor airflows. In this section, the RNG version of the LES model which uses an SGS model based on the Renormalization Group (RNG) theory [22] is employed. Comparisons with the SKE model and its calibrated version for improving the jet spreading rate, SKE-NL from Case 1 are tested in a 3D room with a partition. This room has experimental data by Posner [23] which is used for validation (Figure 7).

The room geometry has a floor area of 91.4 cm x 45.7 cm with a height of 30.5 cm. A partition with a height of 15 cm is located in the middle of the room. The inlet has the same size as the outlet, 10 cm x10 cm.

According to the experimental study of Posner [23], the inlet velocity (U_{inlet}) is 0.235 m/s with a uniform profile. Based on the inlet velocity and the inlet width, the Reynolds number of the inlet airflow is determined to be 1600. The turbulent intensity at the inlet is expected to be very low and assumed to be 1%. To ensure that the solution achieved sufficient statistical independence from the initial state, time-averaged results are obtained from the instantaneous values after the airflow simulation is marched for 2000 non-dimensional time steps, representing 100 seconds in physical time. After this time, the instantaneous values such as the airflow velocities are averaged over 10 000 non-dimensional time steps, or 500 s in physical time.

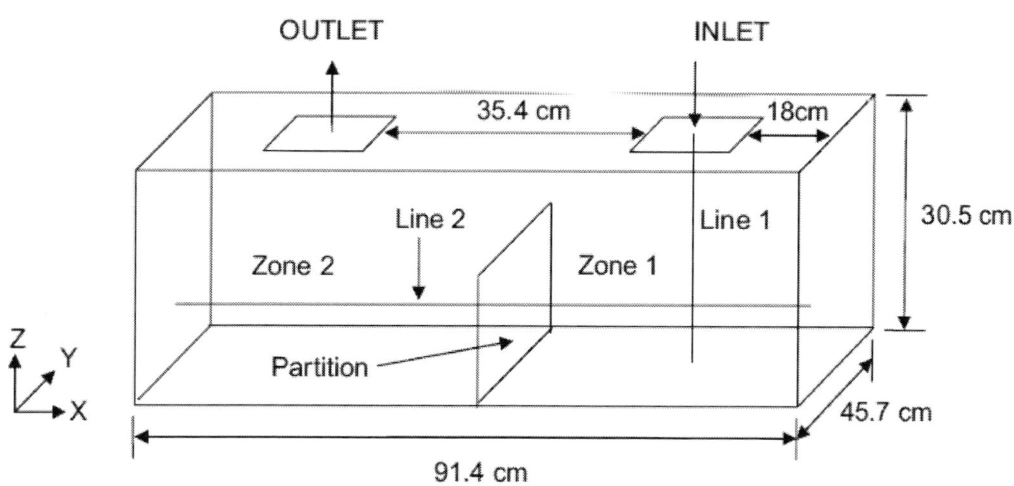

Figure 7. Schematic of the scaled down room with its dimensions used in the experiments by [23]. This room has a partition which is representative of building office layouts that use partitions.

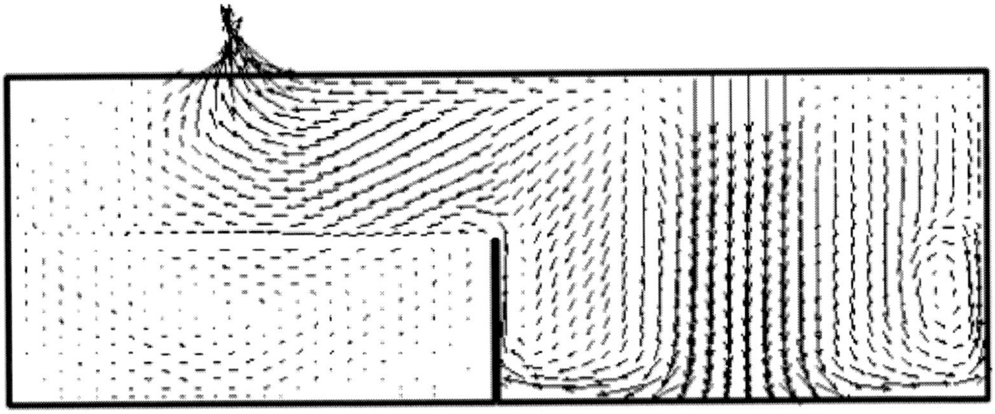

a. LES model simulation.

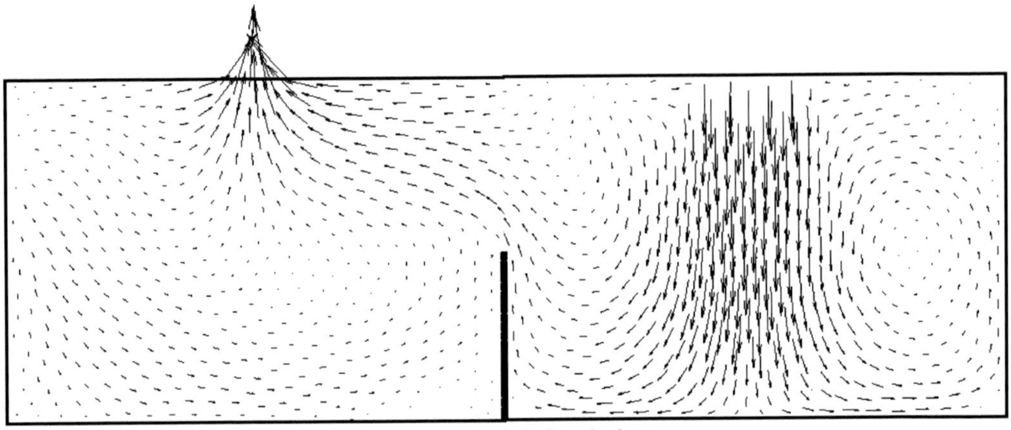

b. k-ε model simulation.

Figure 8. Predicted time-averaged air velocity vector field of the mid-plane (y=0.2285m) for the a) LES model and b) k-ε model simulation.

The simulated time-mean velocity field at the mid-plane (y = 0.2285 m) of the model room is shown in Figure 8. The main features of the flow including the jet stream from the inlet, the recirculation structures in the region next to the right wall (zone 1) and the region beneath the outlet are adequately predicted by the LES model.

The vector plot for SKE model shows a similar general trend to the LES simulation. However in the region near the outlet it does not capture the flow distributing downwards which is found in the LES simulation. The vector plot for the SKE-NL model showed a near replica of the SKE results and therefore is not shown here for brevity. The comparison of the predicted vertical air velocity along the vertical inlet jet axis (line 1 in Figure 7) against the experimental data is presented in Figure 9a. Good agreement is achieved for the LES model, although there is a marginal discrepancy at the end of the jet that could be caused by the numerical diffusion. The SKE and SKE-NL velocity magnitude profiles are predicted well near the jet inlet, but are somewhat underpredicted in the region far from the wall.

Figure 9b shows the comparison between the predicted and measured vertical air velocity (v-velocity) component along the horizontal line at the mid-partition height (line 2 in Figure 7). From the left wall to the partition, the LES model yields a smooth velocity profile that agrees well with the experiment data.

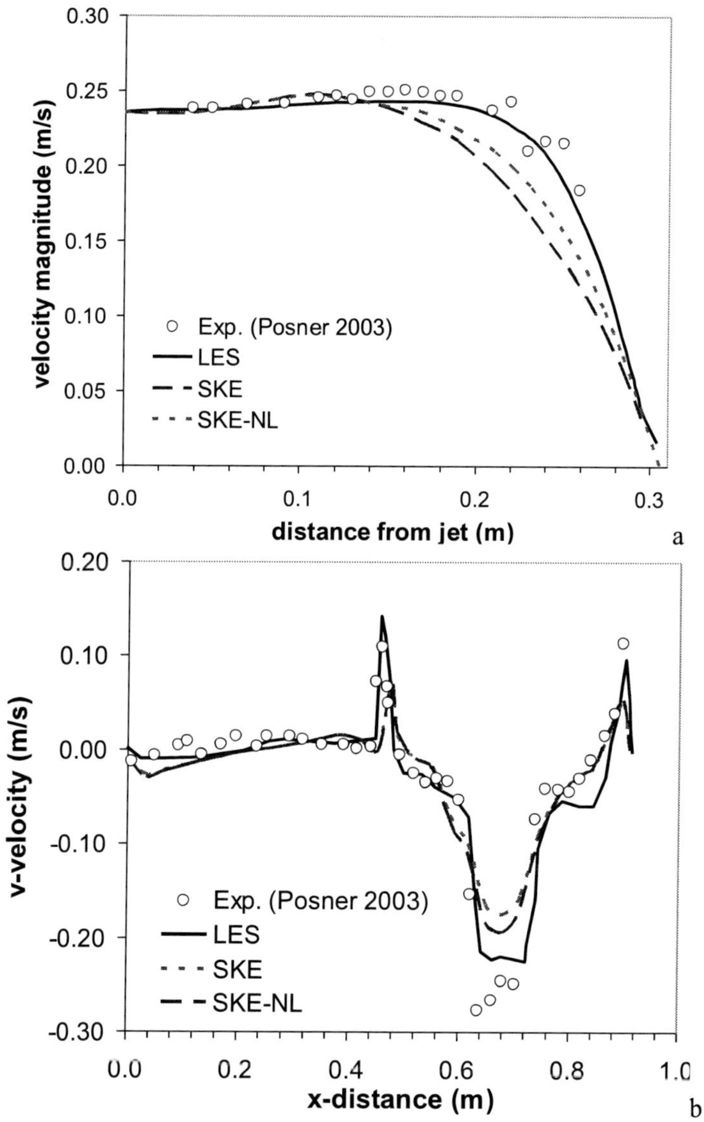

Figure 9. Comparison between simulated and measured velocity components: (a) along the vertical inlet jet axis (line 1) and (b) along the horizontal line at mid-partition height (line 2).

In the regions next to the partition and the right side wall, this LES model successfully captures the highest positive vertical velocities. Generally, the LES model performs very well; good agreement is obtained between the predictions and measured data and the flow trends are successfully captured through the LES model. It is well known that the Smagorinsky SGS LES model, which is based on high Reynolds number dynamics, predicts a non-zero turbulence viscosity in laminar flows. The RNG version of the LES model remedies this unphysical problem by introducing a ramp function. In low turbulent flow regions such as in transitional and laminar flows, the argument of ramp function becomes negative and the ramp function becomes zero. As a result, the effective viscosity recovers the molecular viscosity. This thereby enables the RNG-based SGS eddy viscosity to model the LRN effects in transitional, laminar flows and near-wall regions that have been encountered in indoor airflow. For the SKE and SKE-NL the vertical velocity profiles are predicted well near the left wall in the left side of the partitioned room. The two peaks in the profiles (one near the left wall and one near the right wall) are not captured well. The peak velocity near the left wall is caused by the partition while the peak near the right wall is caused by the nearby wall forcing a recirculation of the flow. A considerable dip in the profile is found underneath the jet location. Here the SKE and SKE-NL model are not able to capture the effect of the jet velocity penetration. Overall it must be noted that the SKE-NL improves the predicted results over the SKE model, despite the fact that the model was tuned under different conditions.

One of the disadvantages of using RANS turbulence models is its time-averaged results. In contrast the LES model can account for the historical and the transport effects on turbulence, i.e. it captures the instantaneous turbulence properties. To illustrate this point, the instantaneous velocity profile at the mid plane is shown in Figure 10 which shows strong mixing of air, and a number of local eddy regions at the partition height in the left side of the partition. In comparison with the averaged velocity vectors in Figure 8, these local eddies are not seen. The lack of instantaneous information can lead to inappropriate evaluation and design of ventilation systems. This has been confirmed through the LES investigation of natural ventilation in a building with a large opening [24][25]. In that study, it was found that fresh air streamlines reached the opposite wall to the opening in the instantaneous prediction, while streamlines in the time averaged prediction produced a recirculation zone which did not penetrate so profoundly into the enclosure. Thus, LES provides a way to account for the instantaneous information.

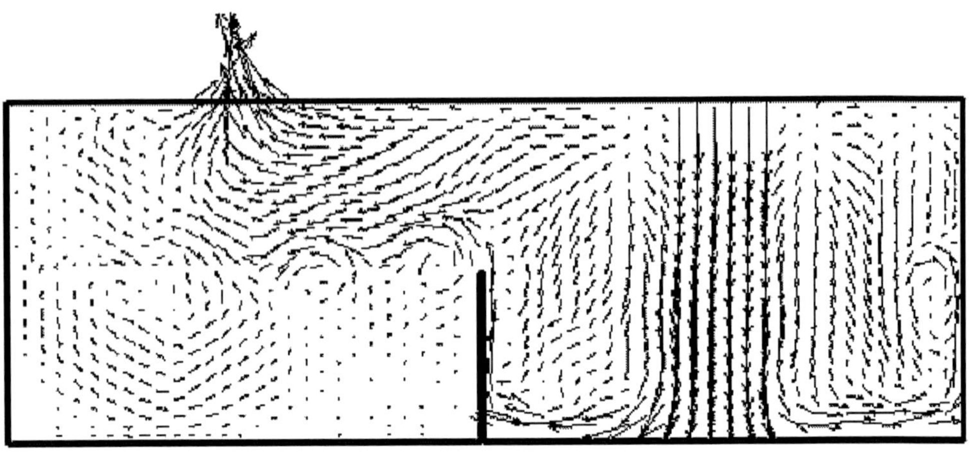

Figure 10. Predicted instantaneous velocity field of the mid-plane in room 1 using LES transient model.

SUMMARY

Turbulence models commonly used in CFD programs were tested for their performance to predict typical indoor airflows. These flows are low velocity in nature and exists in the low-turbulence regime which can cause problems for turbulence models that were historically developed for high turbulence flows. It was found that these models failed to predict the spreading rate of wall jets with lower Reynolds number. This is especially important for room airflows which often exhibit low inlet velocities, leading to transitional effects in the flow regime. For this reason a simple calibration procedure for eddy viscosity turbulence models which ensure the log-law and shear-layer behaviour was presented. It was shown that the calibration of such models improve significantly the predicted velocity distribution in ventilated rooms. The calibrated model was then applied to a different room geometry and compared with experimental and LES modelling. The RNG LES model used excellent predictions for the transitional effects in the velocity field. While the calibrated model did not perform as well as the LES model it did provide improvements over the standard model. Finally the inability of RANS models to provide instantaneous information such as the instantaneous air velocity and turbulence showed that this influence needs to be considered when evaluating designs for ventilation systems.

REFERENCES

[1] Jamiriska, M., L. Morawska, and D.S. Ensor, Control strategies for submicrometer particles indoors: model study of air filtration and ventilation. . *Indoor Air*, 2003. 13: p. 96-105.

[2] Ellis, M.W. and E.H. Mathews, Needs and trends in building and HVAC system design tools. *Building and Environment*, 2002. 37(5): p. 461-470.

[3] Spengler, J.D. and Q. Chen, Indoor air quality factors in designing a healthy building. *Annual Review of Energy and the Environment* 2000. 25: p. 567-600.

[4] Chen, Q. and Z. Zhai, The Use of CFD Tools for Indoor Environmental Design, in Advanced Building Simulation, A. Malkawi and G. Augenbroe, Editors. 2004, Spon Press: New York. p. 119-140.

[5] Zhai, Z., Applications of CFD in Building Design: Aspects and Trends. *Indoor and Built Environment,* 2006. 15(4): p. 305-313.

[6] Kato, S., K. Ito, and S. Murakami, Analysis of visitation frequency through particle tracking method based on LES and model experiment. *Indoor Air,* 2003. 13(2): p. 182-193.

[7] Launder, B.E. and D.B. Spalding, The numerical computation of turbulent flows. *Computer Methods in Applied Mechanics and Engineering*, 1974. 3(2): p. 269-289.

[8] Shih, T.H., et al., A new k-ε eddy viscosity model for high Reynolds number turbulent flows. *Comput. Fluids*, 1995. 24: p. 227–238.

[9] Yakhot, V. and S.A. Orszag, Renormalization group analysis of turbulence. I. Basic theory. *Journal of Scientific Computing,* 1986. 1(1): p. 3-51.

[10] Wilcox, D., Turbulence Modeling for CFD1993, 5354 Palm Drive, La Canada, California 91011: DCW Industries, Inc.

[11] Menter, F.R., Improved two-equation k-w turbulence model for aerodynamic flows, in ASA TM-6 1039751992.

[12] Menter, F.R., Two-equation eddy-viscosity turbulence models for engineering applications. *AIAA J.*, 1994. 8(32): p. 1598-1605.

[13] Durbin, P.A., Near-wall turbulence closure modeling without "damping functions". *Theoretical and Computational Fluid Dynamics,* 1991. 3(1): p. 1-13.

[14] Davidson, L., P.V. Nielsen, and A. Sveningsson, Modification of the v2f model for computing the flow in a 3D wall jet. *Turbulence, Heat and Mass Transfer,* 2003. 4: p. 577-584.

[15] Gibson, M.M. and B.E. Launder, Ground effects on pressure fluctuations in the atmospheric boundary layer. *Journal of Fluid Mechanics*, 1978. 86(03): p. 491-511.

[16] Fu, S., B.E. Launder, and M.A. Leschziner. Modeling Strongly Swirling Recirculating Jet Flow with Reynolds-Stress Transport Closures. in Sixth Symposium on Turbulent Shear Flows. 1987. Toulouse, France.

[17] Launder, B.E., Second-moment closure and its use in modelling turbulent industrial flows. *International Journal for Numerical Methods in Fluids,* 1989. 9(8): p. 963-985.

[18] Launder, B.E. and N. Shima, Second-Moment Closure for the Near-Wall Sublayer: Development and Application. *AIAA Journal,* 1989. 27(10): p. 1319-1325.

[19] Moser, R.D., J. Kim, and N.N. Mansour, Direct numerical simulation of turbulent channel flow up to Ret = 5590. *Physics of fluids*, 1999. 11(4): p. 943-945.

[20] Deo, C., J. Mi, and G.J. Nathan. Experimental investigations of the effect of Reynolds number on a plane jet. in 16th Australasian Fluid Mechanics Conference (AFMC). 2007. Crown Plaza, Gold Coast, Queensland, Australia.

[21] Durbin, P.A. and B.A. Pettersson-Reif, Statsitical Theory and Modeling for Turbulent Flows2001, New York, Weinheim, Brisbane, Singapore, Toronto: John Wiley and Sons, Ltd.

[22] Yakhot, A., et al., Renormalization group formulation of large-eddy simulation. *Journal of Scientific Computing,* 1989. 4: p. 139-158.

[23] Posner, J.D., C.R. Buchanan, and D. Dunn-Rankin, Measurement and prediction of indoor air flow in a model room. *Energy and Buildings,* 2003. 35(5): p. 515-526.

[24] Jiang, Y. and Q. Chen, Study of natural ventilation in buildings by large eddy simulation. *Journal of Wind Engineering and Industrial Aerodynamics*, 2001. 89(13): p. 1155-1178.

Estimated Date of Submission: 5th January 2011

Publication: Nova Publishers Book Chapter in Ventilation (intended)

In: Ventilation: Types, Standards and Problems
Editors: V. A. Romano and A. S. Duval, pp. 173-197
ISBN 978-1-61324-424-1

Chapter 7

An Engineering Design Analysis of a Wind Tower

Ben Richard Hughes
Dubai Energy Research Group,
Heriot-Watt University, Dubai Campus, U.A.E

Abstract

This study investigates the integration of natural ventilation devices into new and existing buildings to meet legislative ventilation requirements. The work focus on the engineering design of a modern wind tower, moreover the individual contribution of each piece of the geometry to the overall effectiveness of the device. Computational Fluid Dynamics (CFD) is used to simulate the air flow and calculate the delivery rates for comparison against occupied space requirements. Presented here for the first time is full scale experimental testing of the device in the natural environment, and the effect of the geometrical variations in an actual building. The experimental work was used to validate the CFD investigations and generate useful performance data as opposed to controlled laboratory conditions testing. The results show that careful component design improves the performance of the device between 27 -45%. Moreover, the wind tower exceeds the minimum legislative requirements at low external wind velocities without the aid of mechanical intervention. Each component is evaluated numerically and experimentally allowing designers and engineers to select and combine to create an individual wind tower to suit the occupied space requirements.

1. Introduction

Ventilation of occupied spaces places a vital role in the health and well-being of both the occupants and the building itself. Well publicised building related illness such as Sick Building Syndrome (SBS) have increased public awareness of the necessity of a fresh air supply to their places of work.

There are two methods of ventilating occupied spaces; natural ventilation and mechanical ventilation. Mechanical ventilation uses mechanical machinery to generate or move air volumes. Typical examples of mechanical ventilation are fans and air conditioning units. Natural ventilation refers to the use of non-mechanical methods to supply fresh air. Typical examples of natural ventilation are windows and doors.

Mechanical ventilation systems require a significant proportion of the buildings energy supply. The energy usage of buildings has, until recently, not been an issue due to energy being in ready supply and relatively cheap to obtain. However, a combination of diminishing energy supplies and International treaties to cut carbon emissions have seen energy prices soar and government legislation increase.

Natural ventilation is a passive method of providing high quality air to buildings without the aid of mechanical systems, thus enabling fresh air delivery to occupants utilising sustainable and energy efficient methods.

Wind driven ventilation is not a modern development. In hot climates, such as Iran, it was used in buildings from as early as 3000BC. The Malgalf or wind catch was used by ancient Egyptians and is represented in wall paintings from the period 900AD [1].

The Malgalf, consists of a shaft rising above the height of the building, which traps the air at higher velocity and channels it down into the interior of the building. This device is used in the hot climates of Arabia. However, modern day equivalents are available for use in new and existing buildings using generic terms such as wind towers.

The device sits on top of the building, trapping the air at higher velocity and delivering it into the interior of the building. The current cross-ventilated design combines the velocity, pressure, and density of air to produce the wind driven ventilation.

Commercial wind vents are divided into quadrants, which allow fresh air to enter as well as stale, used, air to escape irrespective of the prevailing wind direction. There are two driving forces for the wind vent. The primary force provides fresh air driven by the positive air pressure on windward side, exhausting stale air with the assistance of the negative air pressure on the leeward side, shown in Figure 1.

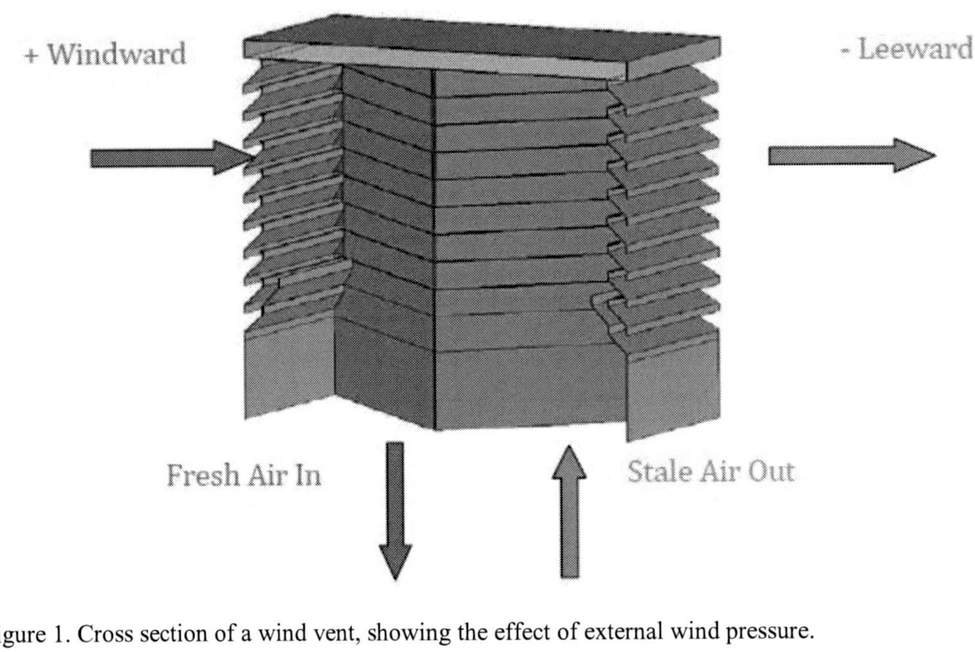

Figure 1. Cross section of a wind vent, showing the effect of external wind pressure.

The main objective of this work is to systematically examine the components of a wind vent and their relative contribution to its effectiveness as a fresh air supply system. Thus enabling engineers to custom design a wind tower to fit the ventilative requirements of specific buildings, either new build or through retro-fit. To achieve this both Computational Fluid Dynamic (CFD) modeling and full scale experimental testing was conducted, with the device installed in a real building and monitored in the natural environment.

1.1. Key Components of a Commercial Wind Tower

To facilitate experimental variables, a standard (commercially available) wind vent, was designed and modified. The individual components were all modified to allow for ease of adjustment, and removal if required. It was envisaged that the overall height of the structure would be reduced (by removing standard blades) to monitor the air-flow present, this was incorporated into the bespoke design. The main components of the overall assembly were the terminal roof (Or top hat), cross- dividers structure (internal), louver blades, internal spacer, control damper unit and egg crate grill (or diffuser). The complete assembly is shown in Figure 2. The internal structure is not visible from figure 2, but it consists of a cross divider and control damper assembly shown in Figure 3.

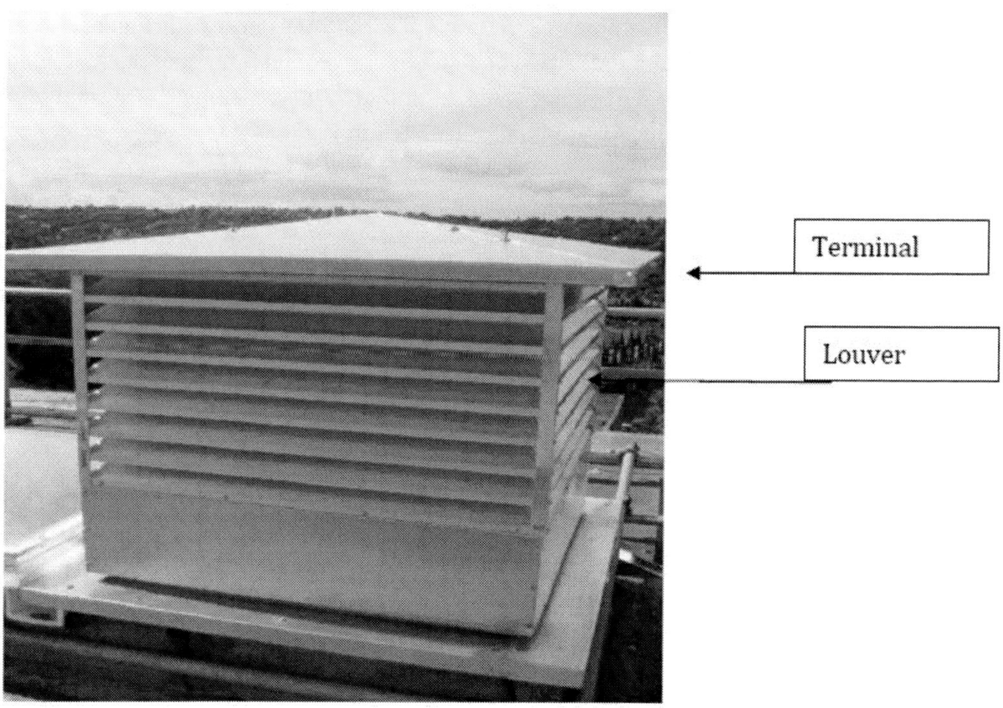

Figure 2. Complete experimental wind vent assembly in position.

Figure 3. Cross divider and control damper assembly.

1.2. Computational Fluid Dynamic Analysis

The CFD modeling carried out in this investigation was through Ansys commercial software using the pre processor Gambit to create the geometry and grid with Fluent used as the processor and post processing the results. The wind tower geometry and the micro-climate (occupied space), control damper unit, spacer, and a simulated macro-climate (external wind velocity domain) were created using Gambit and illustrated in Figure 4.

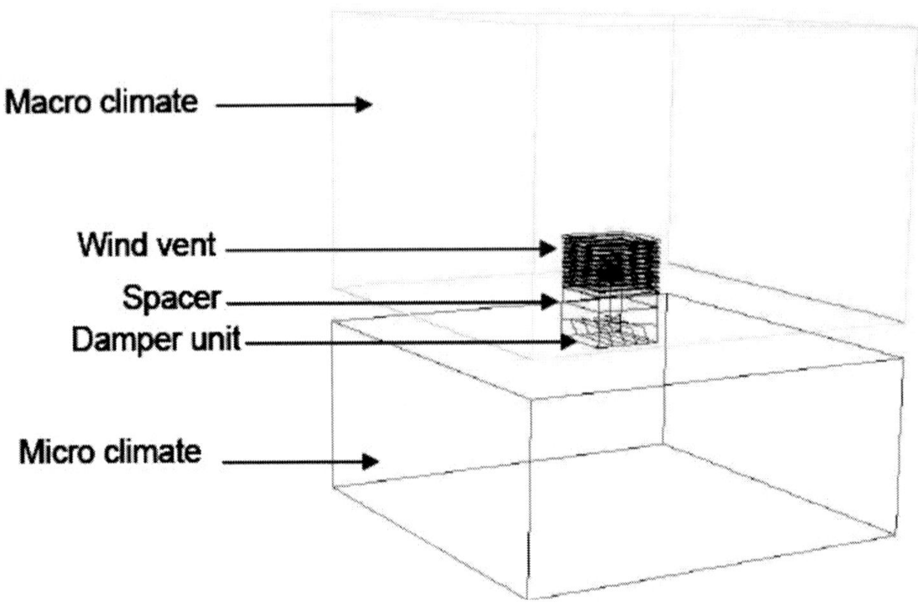

Figure 4. Gamit physical representation of geometry under investigation.

Defining the computational domain (from the physical domain) consists of four key stages, namely defining boundary conditions, creating the volume mesh, examining the mesh and mesh independency (Grid adaption).

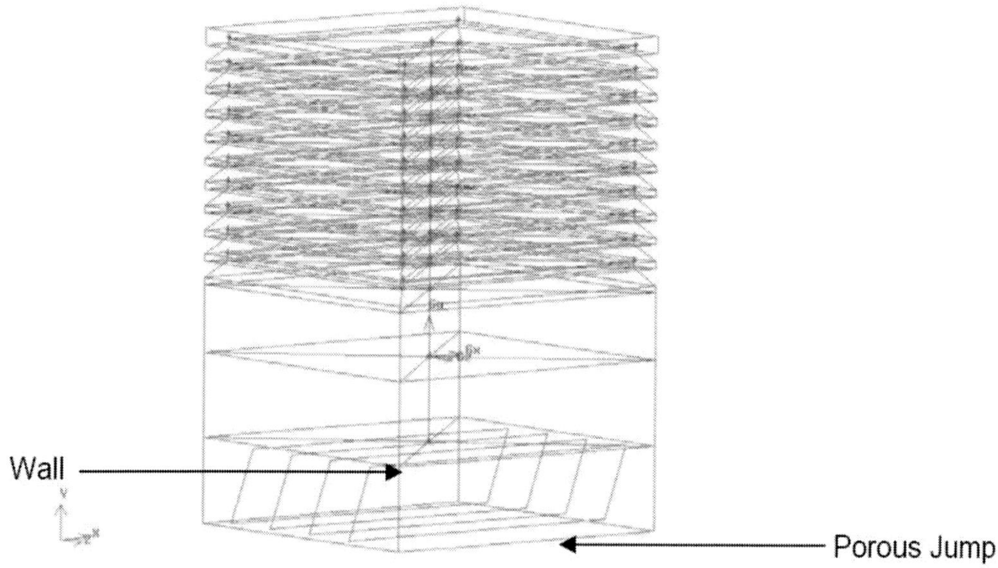

Figure 5. Wind vent fluid volume.

1.2.1. Boundary Conditions

The physical domain contained three fluid volumes, the macro-climate, wind vent and micro-climate. A wall boundary condition was used to clearly define the flow limitations in each region. The macro-climate fluid volume (used to simulate the external velocity flow field) generates a velocity into the wind vent and accepts an exhaust air out-flow from the wind vent.

To generate a velocity flow field, one horizontal plane was named as a velocity inlet, with the opposite horizontal plane named as a pressure outlet. The velocity inlet was set to 4.5 m/s (unless otherwise stated), as this is the UK average wind speed [2]. The remaining planes of the macro-climate were named as symmetry planes - used to model zero shear-slip walls - thus alleviating potential distortion of the macro-climate flow.

The wind vent fluid volume contained two boundary conditions, wall and porous jump. Wall condition was used to simulate any geometrical obstruction to the fluid flow, for example, cross-dividers and dampers. Porous jump was used to simulate the fluid flow through the diffuser (egg crate grill) of the unit shown in Figure 5.

Porous jump acts as a thin membrane (or filter) with user defined inputs of permeability and medium thickness. The commercial wind tower used a 1m square diffuser (evenly spaced mesh) hence from measurements it was deduced that the permeability of the medium was $0.92m^2$. The turbulence modeling was carried out after extensive review of previous works to establish the most appropriate (3,4,5 and 6). The k-ε model has been used in this study with the following equations solved:

- Continuity equation
- Momentum conservation equation
- Navier-Stokes equation
- Kinetic and epsilon transport equation

In addition the defined boundary conditions require value inputs to accurately define the model as depicted in table 1.

Table 1. Boundary conditions requiring input

Boundary condition	Input
Velocity inlet	4.5 m/s
Pressure outlet	Default (atmos)
Porous Jump Permeability	0.915849 m^2
Porous Jump thickness	0.01m

1.2.2. Volume Mesh

The meshing scheme used in this investigation was the Tet/hybrid element combined with the Tgrid type. This scheme specifies that the mesh is composed primarily of tetrahedral elements, and also includes hexahedral, pyramidal, and wedge elements where appropriate. The selection of this scheme was due to the nature of this investigation whereby geometrical variations would be performed; requiring a variation of element types to produce an accurate mesh shown in Figure 6.

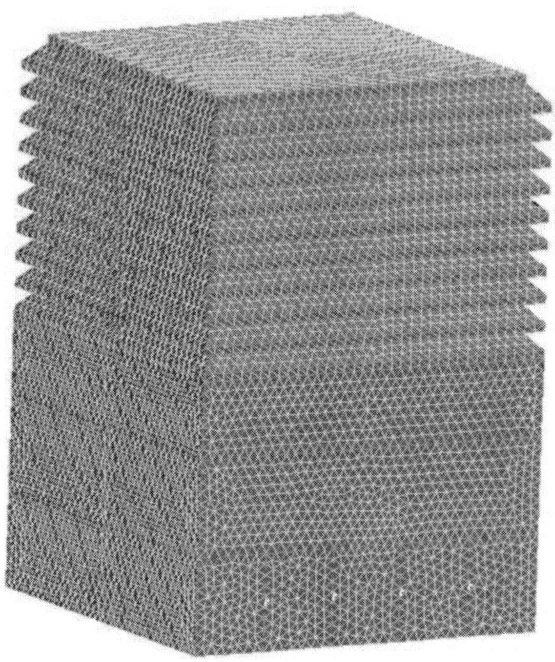

Figure 6. Meshed wind vent geometry.

1.2.3. Grid Adaption

The grid adaptive technique used in this investigation was the *hp* method, (7). The complete computational domain was split and refined in three areas (corresponding to the three fluid volumes): the wind vent and immediate surrounding zone; remaining macro-climate; and finally the micro-climate shown in Figure 7.

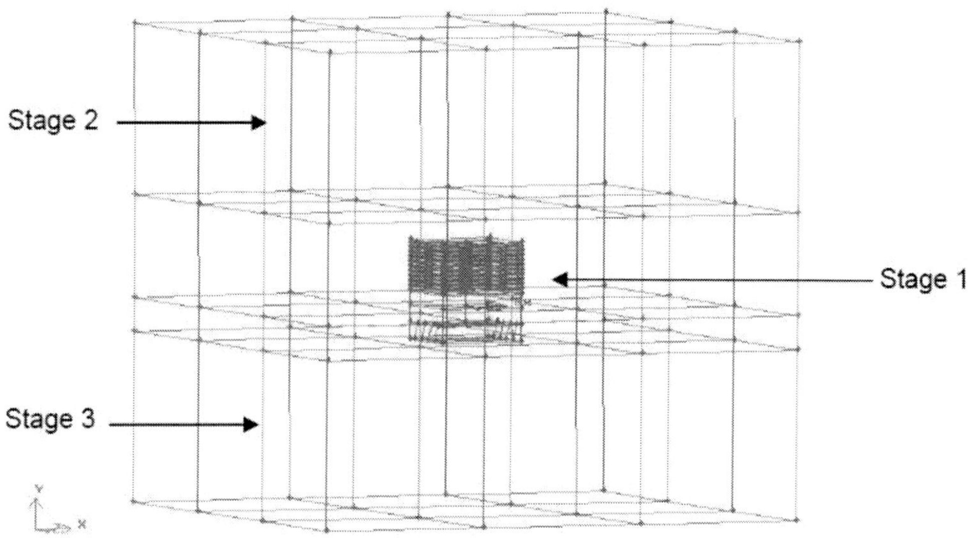

Figure 7. Grid adaption areas (identified by stage of adaption).

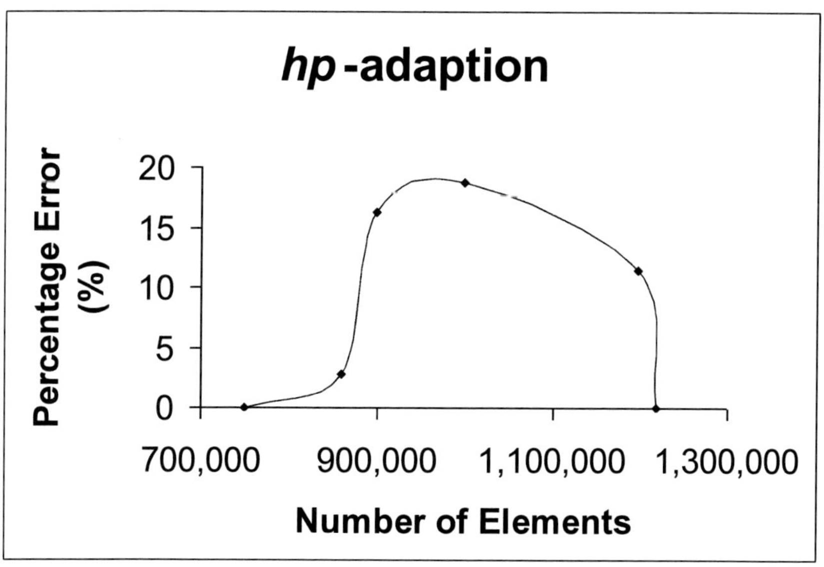

Figure 8. Error reduction through successive *hp*-adaption stages.

The *hp* method required a posteriori error indicator defined as the diffuser average area-weighted velocity result. The grid adaption process increased the number of elements by 445,793. Each stage continued until an acceptable compromise was reached between number of elements, computational time to solve and the posteriori error indication. At 1,196,003 elements the error indication between refinements was at its lowest in the final two stages coupled with the computational time, made it an acceptable compromise. These node spacing's were applied to each investigation, summary shown in Figure 8.

1.2.4. Post Processing

To obtain an accurate picture of the air-flow behaviour through the wind vent and into the micro-climate a series of data points (vertex) were created. From these points, the average vertex velocity (m/s) was obtained and recorded for comparison with experimental results. In total ten data points were created. Nine of the data points were positioned at a height of 1.5 m, with a tenth point centrally positioned at the diffuser entry and exit face. By joining these data points about a 1.5 m high, horizontal plane, velocity and pressure profiles within the micro-climate were obtained, shown in Figure 9.

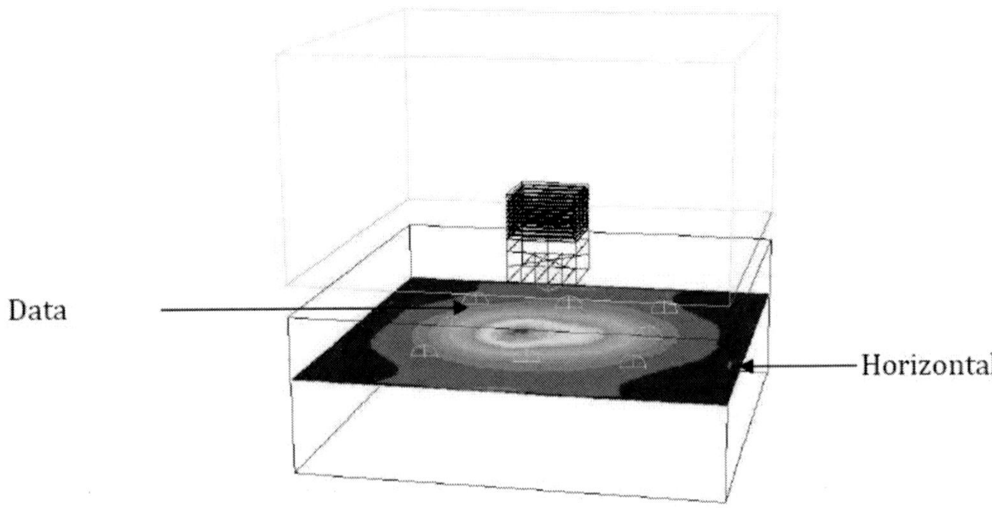

Figure 9. Velocity profile within the micro climate.

1.3. Experimental Analysis

To verify and validate the CFD simulation work carried out as part of this study full scale experimental testing was undertaken. The location of the test facilities was (where possible) matched to the simulation model. The wind vent was a bespoke design (for the purpose of this study) to enable geometrical variation within the overall structure.

The test facility was located within a four-story, flat-roofed building in the Psalter lane campus of Sheffield Hallam University. This area provided an unobstructed air-flow around the building. The flat roof was easily accessible, which facilitated the variation of the terminal geometry. The building stands 15 metres high, with a clearance of approximately 2 metres from the closest surrounding buildings.

To record the climatic data (surrounding the terminal), during the experimental testing, a multi-test probe attachment of the weather station was placed on the roof, with a data logger located inside the building using wireless connections. The test terminal was designed to facilitate multiple geometrical variations. In order to maintain rigidity the internal structure was strengthened. The cross-dividers were encased by a stainless steel frame fixed in position by both rivets and weld, shown Figure 10.

Figure 10. Cross divider structure.

The external louvers were not individually attached to the cross-divider structure. Each louver slide over the structure and connected to the other louvers via external stainless steel spacing bars, using four, 50mm cross thread screws. The overall dimensions of the structure were 1000 mm x 1000 mm with a height of 900 mm.

Two types of louver blades were manufactured. The louvers were distinguished by the pitch angle of the extrusion (with respect to the vertical edge from the horizontal), 35° and 45° respectively. Each louver had a height of 100mm and a depth of 50 mm. Nine louvers made each set, with two sets for experimental testing.

The completed assembly of: the top hat; louver section; and cross divider structure, had a layer of wire mesh (20 mm spacing) around the inside to prevent any birds or small animals entering. Additionally, an aluminium flashing was used to seal the joint between the wind vent and spacer.

1.3.1. Internal Measurements

The micro-climate was selected to most accurately reflect the simulation models. The simulation model had room dimensions of 6 x 6 x 2.5 m, with a volume of $90m^3$. The wind vent was located centrally within the micro-climate. A test grid - marked on the floor - to locate the position of the nine sample points required (in comparison with the simulation model vertex). Samples were taken using a hot wire thermal velocity probe and a portable multimeter. The probe attached to a tripod to locate consistently at the same height and location on the grid to ensure repeatability shown in figure 11.

To record the climatic data of both the macro and micro-climate (in synchronisation) a multi-probe data recorder was placed within the micro-climate.

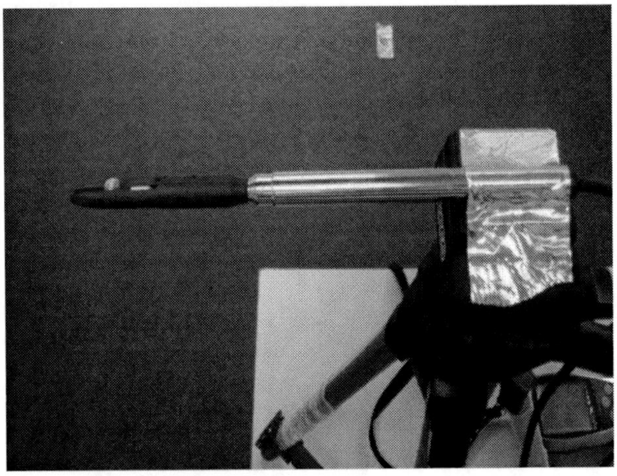

Figure 11. Velocity probe connected to tripod and test meter.

This was located on a wall and recorded all relevant climatic data such as: temperature, wind speed, rainfall, humidity, and daylight hours, and stored it prior to weekly download. Using this procedure the internal measurements were compare to the external climatic data over a period of 3 months using various geometrical configurations.

1.4. Validation

To validate the CFD simulations the computed results are compared with the experimental data, to ensure accurate analysis. Multiple geometrical configurations were carried out which required extensive validation for each set-up. Figure 12 shows a typical validation comparison from this investigation.

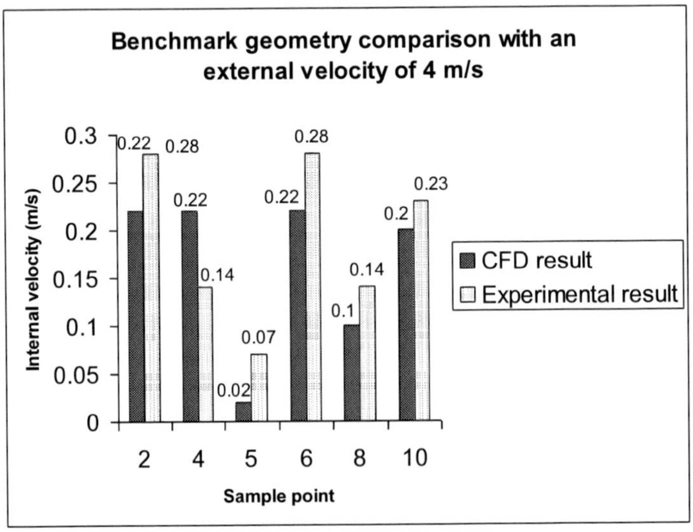

Figure 12. Comparison of the benchmark geometry CFD results against experimental data with an external velocity of 4 m/s.

The error range across the sample points was 3-8%, using such low internal velocities this was as anticipated. Moreover the trend of the sample points was in agreement.

1.5. Results

Using the previously described CFD and experimental techniques the geometry of the wind tower was thoroughly examined to obtain useful data about the design and operation.

1.5.1. Number of External Louvers

To investigate the effect of the number of external louvers on the wind vent four experiments were carried out using two, four, six and eight louvers independently. To achieve this the unused louvers were sealed using a blanking piece shown in Figure 13.

Each experimental set up was monitored against external wind conditions for a period of seven days. Using the device in the natural environment meant that it was not possible to take direct comparison in all cases as the external wind varied, for example using six louvers the external wind was on average 4 m/s, however the week of testing using four external louvers the external wind was 5 m/s. However the CFD modeling was able to use controlled conditions and thus comparisons can be drawn more readily.

Figure 13. Experimental set up using blanking piece to vary the amount of opening external louvers.

Table 2. Simulation results of varying the number of external louvers

Number of external louvers	Diffuser face velocity (m/s)	Diffuser face pressure (Pa)	Internal air distribution (m/s)
2	0.22	0.03	0.31
4	0.42	0.13	0.35
6	0.52	0.21	0.34
8	0.57	0.25	0.34

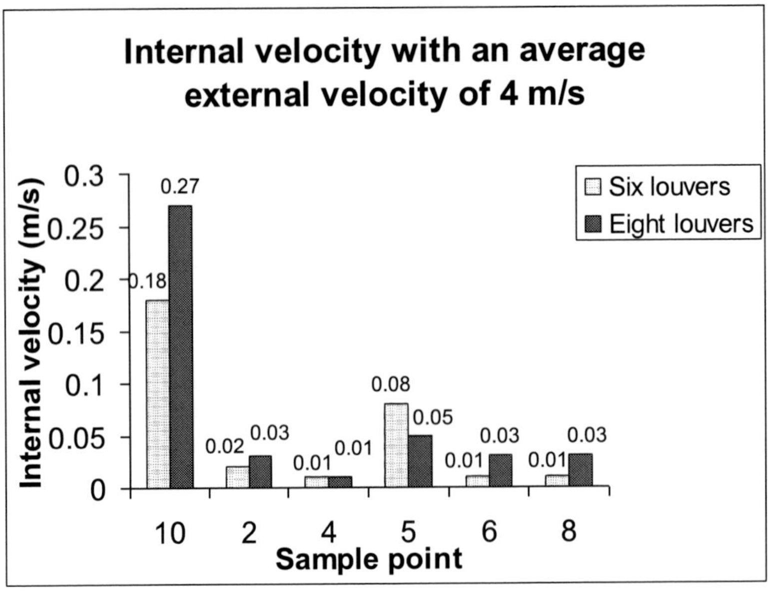

Figure 14. Comparison of the six louver experimental set up with the benchmark case to demonstrate the effect of varying the amount of external louvers.

Shown in Figure 14 is the comparison between eight and six louvers as during these experiments the average external velocity was at 4 m/s.

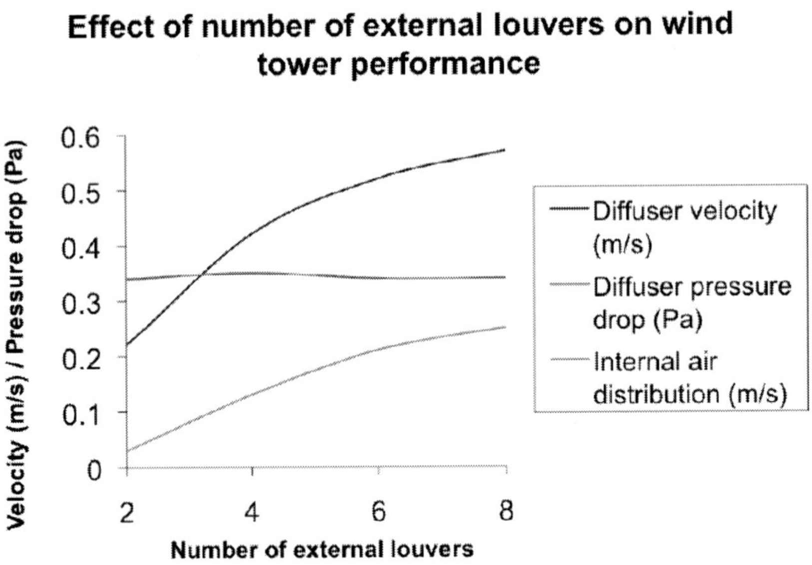

Figure 15. Graph of number of external louvers against velocity and pressure parameters.

From the figure it is evident that reducing the number of louvers, has a reduction in velocity across most sample points. Sample point 10 being the most significant as this is directly below the diffuser into the micro climate, and hence where the air exchange takes place. For quantification of the amount of reduction using reduced louvers the CFD analysis presents clearer definition, as it is run under controlled conditions. The CFD models were run using constant pressure and external velocity of 4.5m/s. Table 2 shows the increase in diffuser face velocity (equivalent to sample point 10) when increasing the number of external louvers. From the results; increasing the number of external louvers leads to a direct increase in the diffuser face velocity and dynamic pressure. This can be shown more clearly in Figure 15.

The internal air distribution remains consistant through the increase in external livers, this leads to the conclusion that the rate of air exchange correlates to the external louvers configuration but the circulation inside the micro climate does not.

1.5.2. Distance between External Louvers

To investigate the effect of varying the distance between the external louvers, three different experimental set-ups were used. The three set-ups were achieved by using spacer bars at the four corners of the wind vent (described previously). Reducing the distance between the louvers had an undesirable effect of reducing the overall height of the louver section, Hence, the internal blanking piece was used at the top louver to alleviate any additional air flow. The three louver spacing's tested were: 20, 30, 50 and 60mm spacings. The experimental set-up can be seen in Figure 16.

Figure 16. Experimental set up for the investigation into the effect of varying the spacing between external louvers, 20mm spacing shown.

In the experimental investigations two case's, namely 30 and 20mm external louver spacing, share the same average external velocity. A direct comparison between the two is shown in Figure 17.

The reduction in spacing from 30 to 20mm had little overall effect on the internal velocity. Sample point ten, showed a reduced velocity which indicated a reduction in air-exchange within the wind vent. The results were as expected, that the effect of reducing the spacing is to reduce the opening area of the wind vent which in turn restricts the air-flow within it. The CFD investigations provided more detailed information about the effect of louver spacing. Five simulation models were solved with an external wind velocity on 4.5 m/s, all other geometry remained constant. Full results are shown in Table 3.

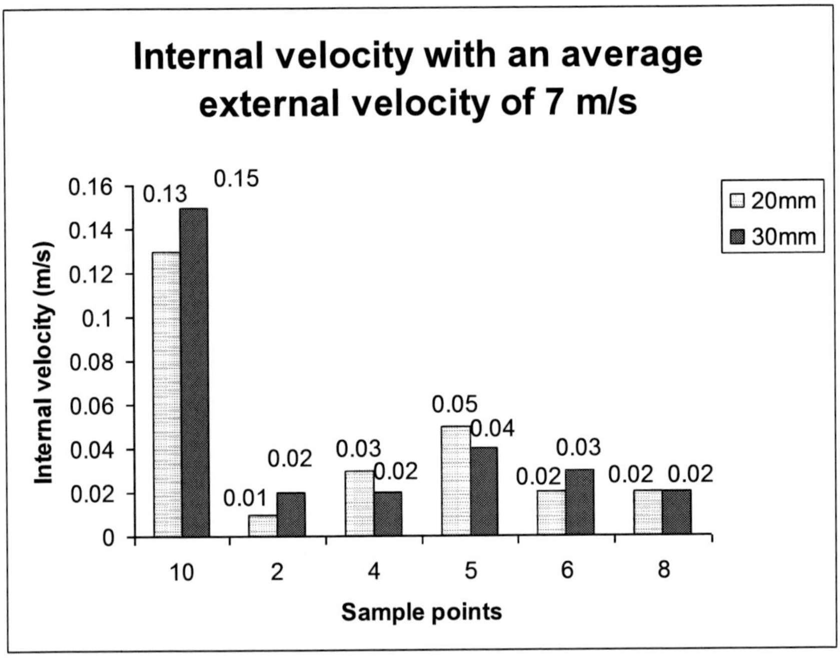

Figure 17. Comparison of the 30 and 20mm external louver spacing experimental set up.

Table 3. Simulation results of the external louver spacing variation

Louver spacing distance (mm)	Diffuser face velocity (m/s)	Diffuser face pressure (Pa)	Internal air distribution (m/s)
10	0.2	0.03	0.34
20	0.38	0.11	0.36
30	0.42	0.15	0.37
40	0.59	0.41	0.36
50	0.54	0.23	0.31
60	0.67	0.42	0.38

From the results it is evident that as the louver spacing increased, the diffuser velocity and pressure also increase. At the 50mm spacing there was an anomaly within the data, (both the velocity and pressure decrease from the 40mm case).

Comparison of the 60 and 40mm case showed a similar diffuser pressure with a higher diffuser velocity in the 60mm case. This comparison showed that although the air-exchange rate had increased, the pressure (driving the flow through the wind vent) had not.

The internal air distribution similarly increased, in proportion to the increased louver spacing. This was as expected that with a greater inlet and outlet area, there is a greater suction effect within the micro-climate. A comparison of the three performance indicators is shown in Figure 18.

The velocity and pressure trends have two peaks, one at 40mm the other at 60mm. The internal air movement similarly has two peaks, one at 30mm, the other at 60mm. Comparing Table 3 and figure 18 the results show that the optimum louver spacing is at 60mm due to the combination of performance indicators being at the highest level at this geometrical configuration.

However, the purpose of the external louver is to deflect prevailing rain whilst allowing air to enter. Therefore, increasing the louver spacing would (in logical terms) have a detrimental effect on deflecting the rain ingress. Therefore, to determine an optimum setting for the spacing it must generate the maximum internal air movement rate (Or suction within the micro-climate) with the minimum louver spacing (to prevent rain ingress). The optimum louver spacing distance for this application is 30mm, further analysis of the velocity contours confirmed this shown in Figure 19.

Using an external louver spacing distance of 30mm theair exchange rate is lower than the 60mm, with a 30% reduction. However, the internal air movement rate has increased by a comparative margin, 27%. The increased benefit to the wind tower is that the smaller louver spacing should aid the louver design effect of minimising rain ingress.

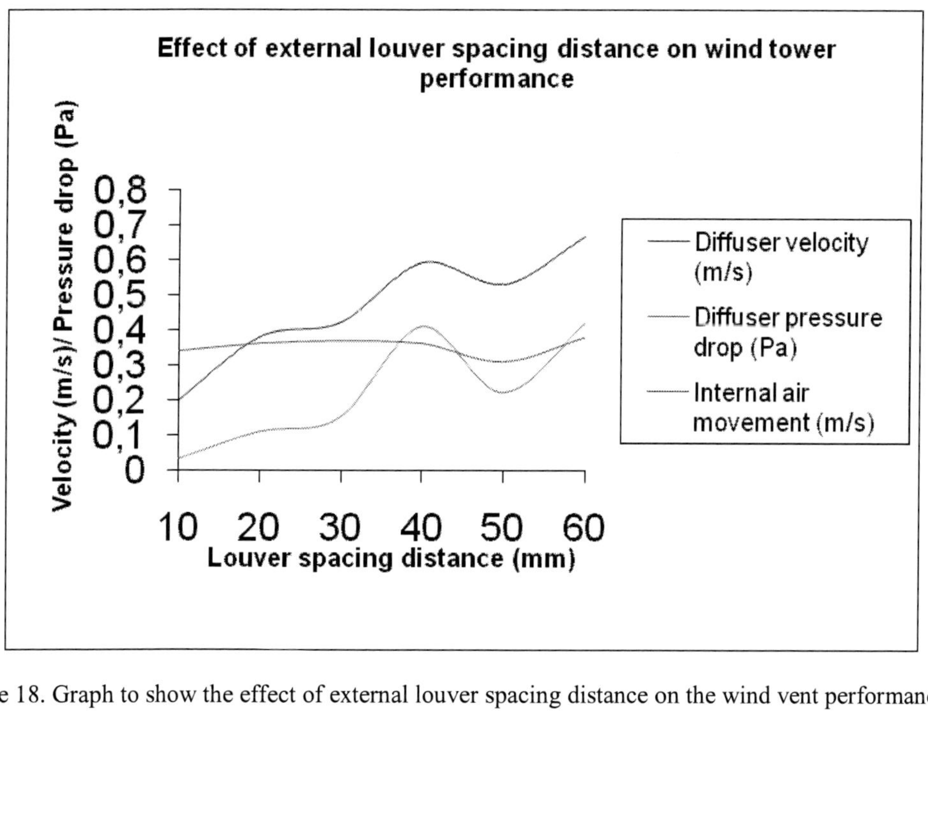

Figure 18. Graph to show the effect of external louver spacing distance on the wind vent performance.

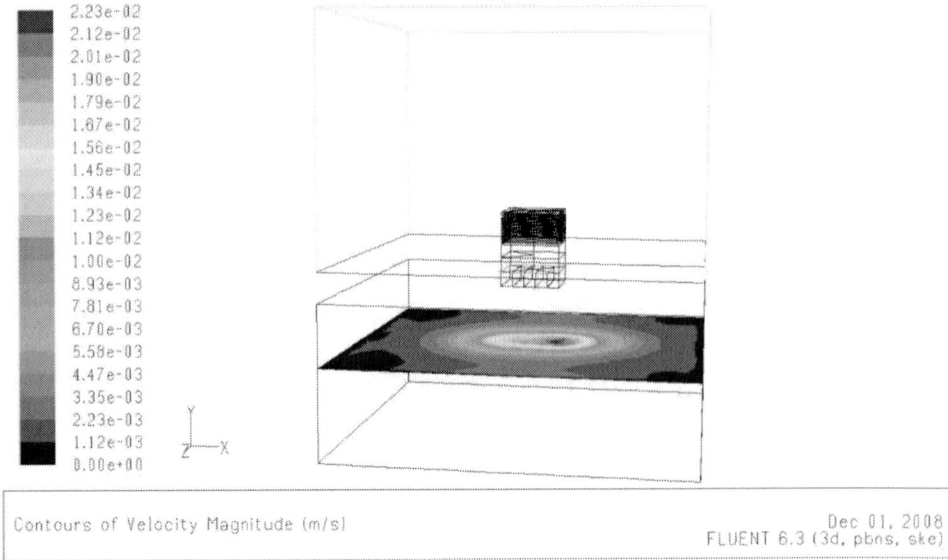

Figure 19. Velocity contours within the micro climate with an external louver spacing distance of 30mm.

1.5.3. External Louver Angle

To investigate the effect of the external louver angle the CFD simulations were carried out first. The reason for this was the cost of manufacture, using CFD it was determined the most effective external angle, and then validated through the experimental testing.

Eight models were solved with an external wind speed of, 4.5 m/s, external louver angles ranged from 10-45 degree. The results are shown in Table 4.

The tabulated results showed that at an external louver angle of 40°, the recorded velocity and pressure across the diffuser, show anomalies in comparison to the other simulation models.

The effect of altering the external louver angle was to cause lift and drag on the airflow around the louvers, similar to aerodynamic theory of airfoils. Thus at an external louver angle of 40 degrees, the flow within the wind tower had actually stalled and the results were unreliable [9].

Table 4. Results for external louver angle simulations

Blade Angle	Diffuser pressure (Pa)	Diffuser velocity (m/s)	Internal air distribution (m/s)
10	4.39	1.89	0.37
15	2.67	1.69	0.34
20	1.67	1.38	0.35
25	1.66	1.42	0.36
30	1.18	1.22	0.36
35	1.29	1.29	0.45
40	5.57	2.84	0.35
45	2.88	1.99	0.31

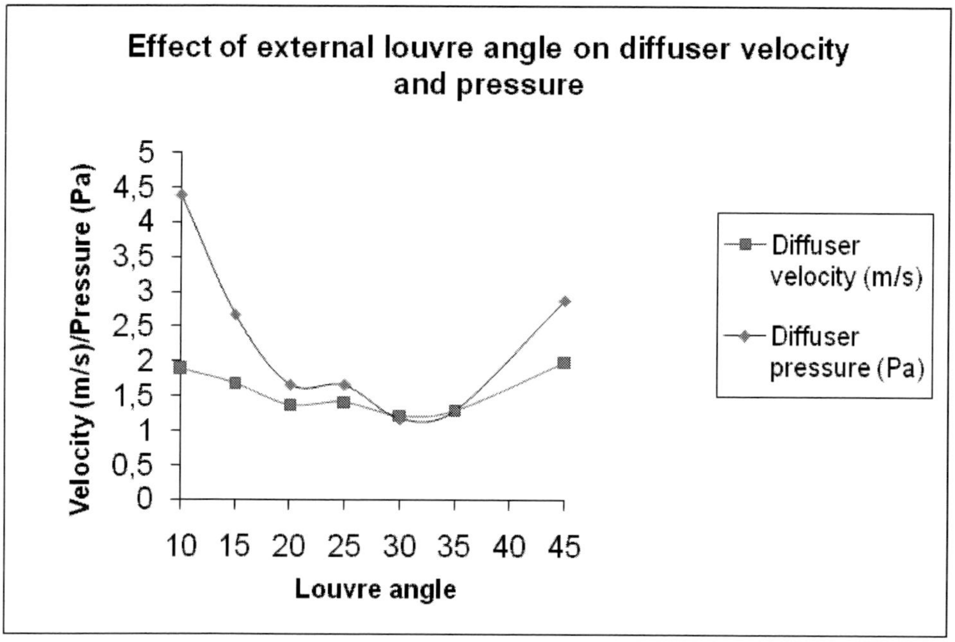

Figure 20. Graph of the effect of external louver angle variation on diffuser velocity and pressure.

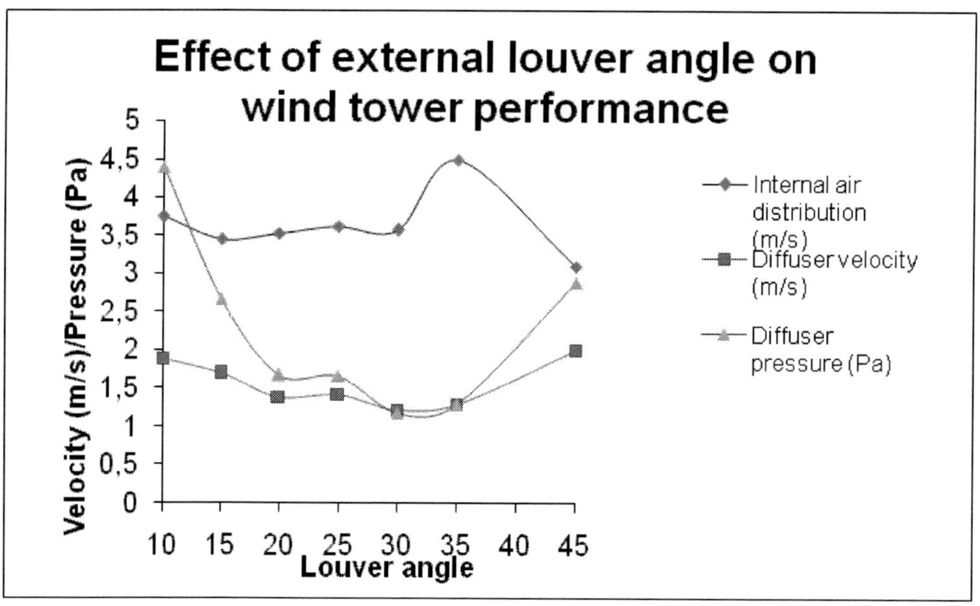

Figure 21. Graph of the effect of external louver angle on wind vent performance.

Comparison of the average diffuser velocity, and dynamic pressure, showed that the external louver angle was insignificant as shown in Figure 20. The graph shows two points of intersection were the 30° case (nearest simulation model), and 35° respectively. The intersections demonstrated the pressure which is equated to flow direction and velocity which is equated to air-exchange working in tandem, thus representing potential for an optimised external louver angle.

A third performance parameter - the internal air movement rate - showed little improvement through the incremental louver angle simulations. However, air movement rate when compared to the louver angle in conjunction with the two remaining parameters demonstrated clear advantage at the 35° intersection, shown in Figure 21.

The wind tower with an external louver angle of 35° shows the internal air distribution has a dramatic increase over the other seven cases. The air distribution velocity has increased by 45% over that of the 45 degree model. For this reason lover angles of 35 and 45 degree were selected and manufactured for experimental investigation.

Figure 22. Experimental wind tower with 35° louvers in place.

A set of eight external louvers were manufactured with an external louver angle of 35° and attached to the wind vent, shown in figure 22. All other geometry remained the same as the benchmark including the louver spacing.

The external louver angle experimental set-up was monitored for five days with the averaged experimental results shown in Table 5.

Table 5. Averaged results for the external louver angle experimental set up

Sample point	External velocity (m/s)	Internal velocity (m/s)
10	3.5	0.27
2	3.6	0.02
4	3.6	0.05
5	3.1	0.05
6	3.2	0.04
8	2.5	0.04

The averaged external wind speed across the five days was 3 m/s. The results are represented in figure 23.

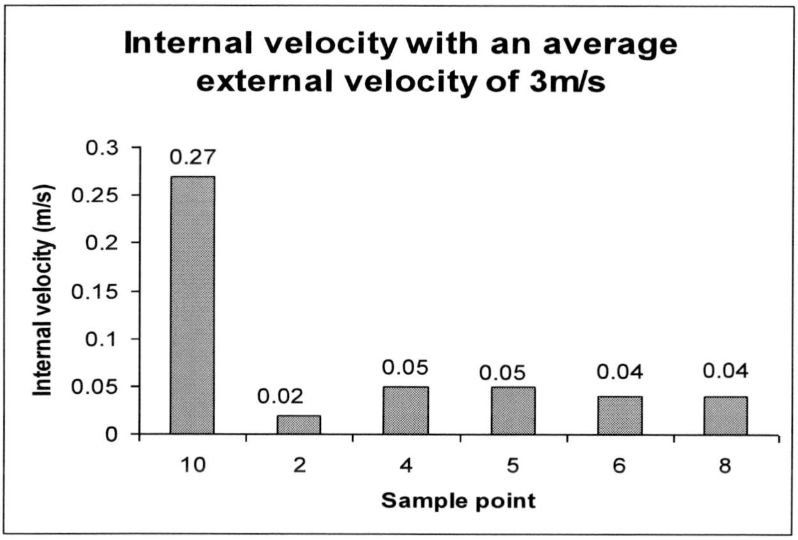

Figure 23. Average internal velocity results of the external louver angle experimental set up.

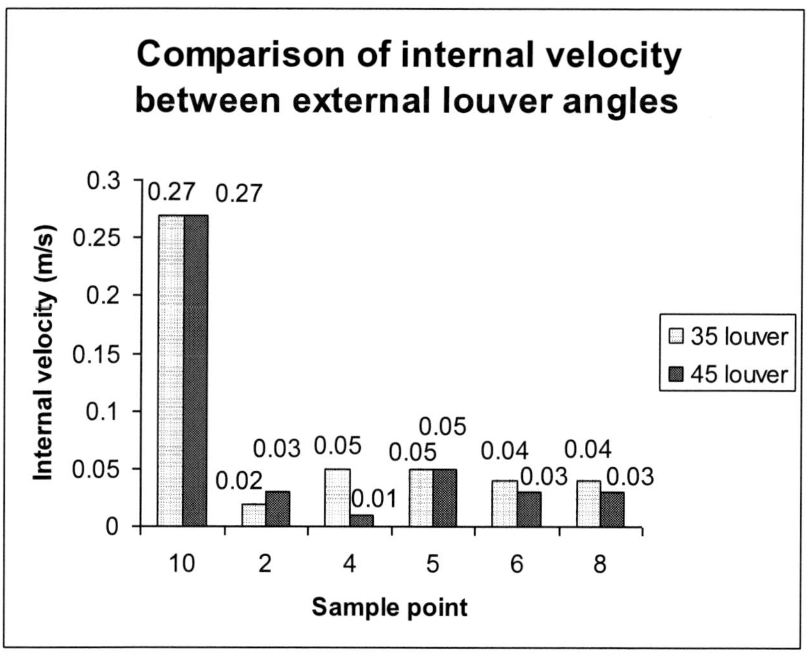

Figure 24. Comparison of the 35° louver angle with the benchmark case.

From the results the external louver angle at 35° has matched the performance of the 45° louvers, shown in Figure 24. Both sample points ten and five are equal in each experimental set-up with 0.27m/s at point ten and 0.05 m/s at point five. However, the average external wind speed is lower in the 35° experiment, with an average of 3 m/s compared to 4 m/s of the benchmark case. Therefore, the 35° louver increased the internal velocity, in comparison to the benchmark. This result concurs with that of the CFD simulation.

1.5.4. Optimum Geometrical Configuration

The optimum geometrical configuration was devised from the CFD simulations and experimental work. The optimum wind tower consisted of four external louvers with 30mm spacing's and a louver angle of 35°. The optimum experimental configuration is shown in Figure 25 and was replicated for CFD simulation.

Figure 25. Experimental set up of the optimum configuration, devised form the previous experimental investigations.

The optimum geometrical configuration set-up was monitored for five days with the averaged experimental results shown in Table 6.

Table 6. Averaged results for the optimum configuration experimental set up

Sample point	External velocity (m/s)	Internal velocity (m/s)
10	5.2	0.24
2	2.9	0.01
4	3.8	0.02
5	3.5	0.05
6	3.9	0.02
8	3.3	0.02

The average external wind speed across the five days was 4 m/s, the results are represented in Figure 26.

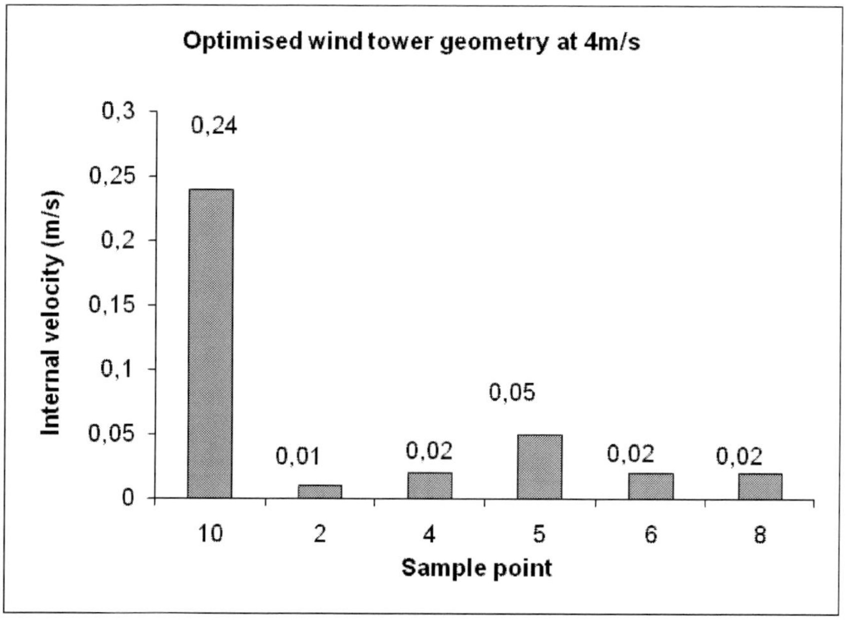

Figure 26. Average internal velocity results of the optimum geometrical configuration experimental set up.

The external velocity in the optimum case was 4m/s. To establish a comparison with the previous CFD and experimental work presented here a second configuration (A benchmark) using eight louvers, with 45 degree external angle and 50mm spacings was set-up and monitored. This set-up was monitored for a sustained period until external wind conditions allowed a comparison between it and the optimum case. This comparison is shown in Figure 27.

From Figure 27 it is evident that the optimised geometry configuration did not out perform the second benchmark configuration. The results showed a marginal advantage in the internal velocity produced by the benchmark in the range of 11 - 20%. However, this range must be considered in the context of the small velocities experienced, the maximum velocity deviation at any sample point was 0.02 m/s.

The experimental results showed that singular geometrical variation improves the wind vent performance. Applying the geometrical variations in conjunction did not produce the expected increase in performance. However, the experimentally optimised geometry achieved a similar performance to that of the benchmark wind vent whilst reducing the number of components and materials used in its construction.

Following the experimental investigations where key external geometrical optimisation has been identified two final simulation models were built to investigate their combined effects on the wind vents performance. The external geometry was selected to be a replica of that used in experimental testing of both optimum and benchmark cases.

The optimum configuration and benchmark cases were solved for a range of external wind velocities, namely 1, 2, 3, 4, 4.5, and 5 m/s. The full results are shown in Tables 7 and 8.

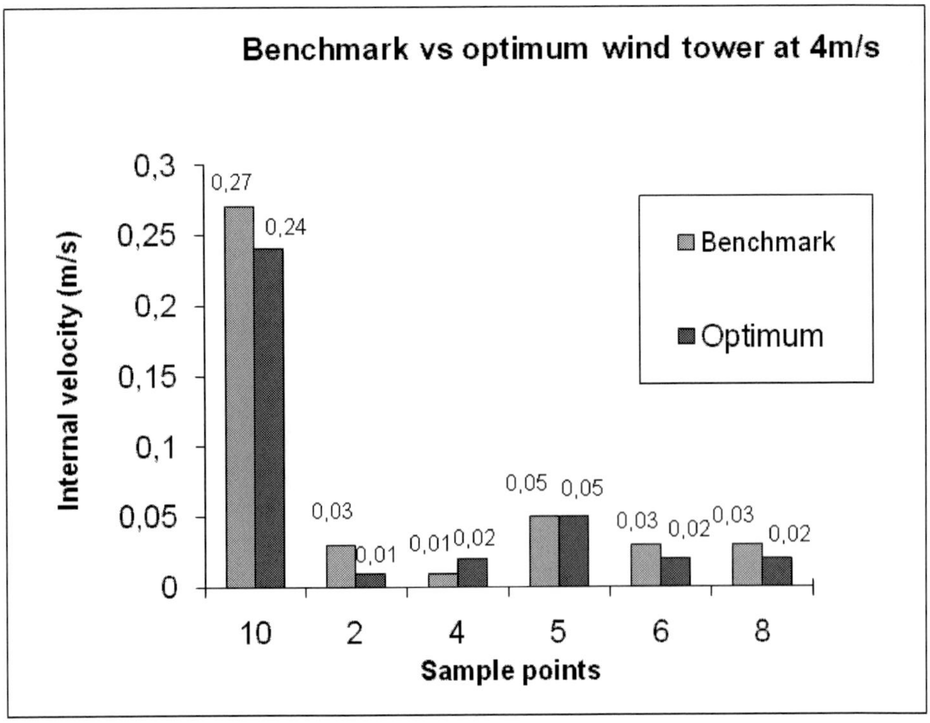

Figure 27. Comparison of the optimised wind tower geometry with benchmark case at an average external wind speed of 4 m/s.

Table 7. Simulation results for optimum external geometry configuration

Velocity inlet speed (m/s)	Diffuser face velocity (m/s)	Diffuser face pressure (Pa)	Internal air distribution (m/s)
1	0.07	0	0.07
2	0.14	0.02	0.14
3	0.2	0.03	0.22
4	0.28	0.06	0.3
4.5	0.32	0.08	0.33
5	0.37	0.1	0.37

Table 8. Simulation results of benchmark geometry

Velocity inlet speed (m/s)	Diffuser velocity (m/s)	Diffuser pressure (Pa)	Internal air distribution (m/s)
1	0.09	0.01	0.07
2	0.21	0.04	0.14
3	0.34	0.09	0.20
4	0.47	0.17	0.27
4.5	0.54	0.23	0.31
5	0.65	0.32	0.34

The results showed that both velocity and pressure were significantly lower in the optimum configuration compared to the benchmark. The velocity was 43% lower at the maximum external wind speed simulated (5 m/s). The pressure was 70% lower at the same external wind speed (5m/s) shown in Figure 28.

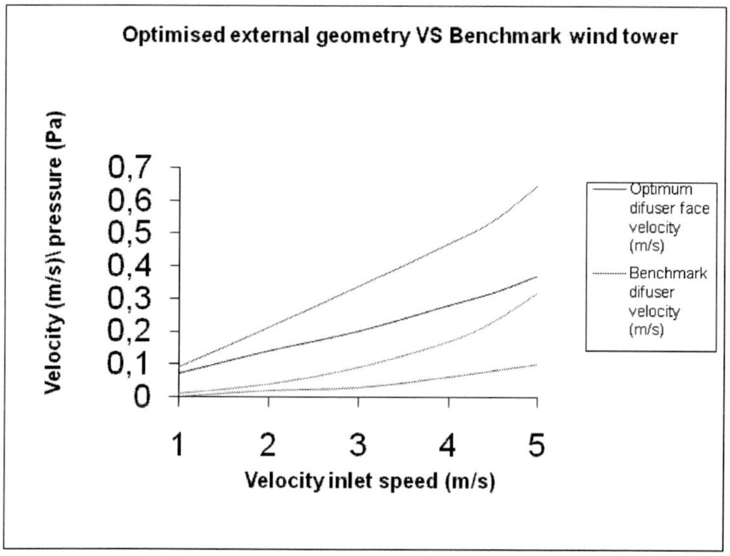

Figure 28. Graph to compare the optimised wind vent performance with the bench mark wind tower.

The optimum configuration provided less air- exchange between macro and micro-climate. The internal air distribution rate for the optimum case is similar to the benchmark with a nominal improvement. The optimum geometry gave a 10% increase in internal air distribution at the highest external velocity simulated (5 m/s) in comparison to the bench mark case, shown in Figure 29.

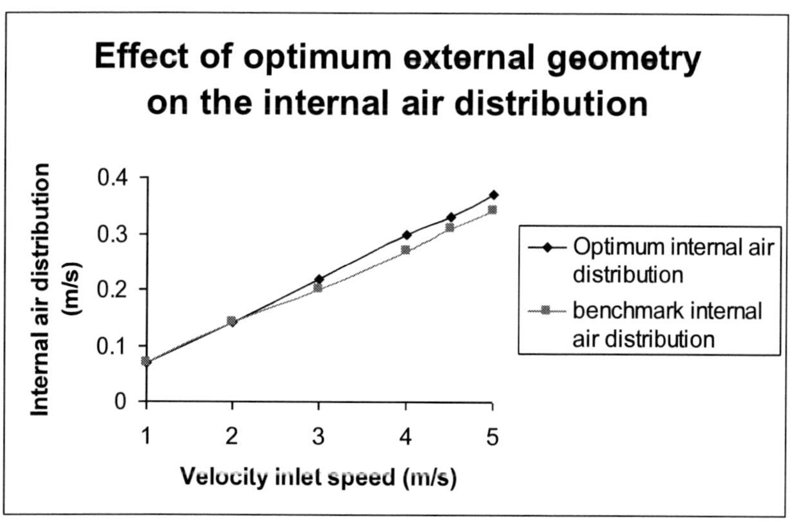

Figure 29. Graph to compare the optimum external geometry with the benchmark internal air distribution.

The optimum case provided greater air distribution within the room. This was as expected as previous investigations showed the suction effect had moved closer to micro-climate. The results showed that the optimum external geometry, when combined, did not represent the optimum wind tower performance in terms of air exchange.

However, this investigation has shown that the internal air distribution rate was marginally improved by optimisation of the external geometry, using 50% less material for the wind tower construction (50% less external louvers).

1.6. Summary

47 CFD simulation models were built incorporating the full geometry of the wind tower. The geometry of the wind vent was systematically examined with the contribution of each section of geometry to the performance of the device, established. The effect of geometrical variation on the flow and air movement from the wind vent has been demonstrated.

This investigation has shown that the performance of the wind vent can be optimised through geometrical variation and the optimum geometrical variation was identified for each section of the wind vent in isolation. The optimum geometrical variations when combined do not provide the optimum wind vent performance.

Extensive far-field experimental data was provided for the wind vent, with each geometrical variation experimentally tested. From the results the key geometrical components have been identified as the external louver angle, distance between the louvers and the number of external louvers.

An optimised wind tower has been numerically and experimentally investigated. The wind tower has shown a similar - but not increased - performance to the benchmark case. This investigation has shown that the optimum components configuration when investigated in conjunction does not provide the optimum performance.

Thus the conclusion of this study I that engineers should first define the air exchange required to meet current legislation. Following this the wind tower can be custom designed using the key geometry defined to meet new build or retro-fit requirements.

Using the techniques described in this chapter it is envisaged that wind towers can become more accessible to architects and designers. The ability to custom build wind towers for any given application provides a platform for integrating these device on a larger scale. Thus the overall building performance in terms of energy consumption from ventilative devices may be improved.

REFERENCES

[1] Santamouris, M. (2002). *Natural ventilation in buildings A design handbook.* United Kingdom, Cromwell Press.

[2] British Standards Institution 1991. BS5925:1991: *Ventilation principles and designing for natural ventilation.* London, British Standards Institution.

[3] Ayad, Samir S. (1999). Computational study of natural ventilation. *Journal of wind engineering and industrial aerodynamics,* 82 (1-3), 49-68.

[4] LI, Liu And MAK, C. M. (2007). The assessment of the performance of a windcatcher system using computational fluid dynamics. *Building and environment,* 42 (3), 1135-1141.

[5] Drori, U. And Ziskind, G. (2004). Induced ventilation of a one-story real-size building. *Energy and buildings,* 36 881-890.

[6] Elmualim, Abbas A. (2006). Verification of design calculations of a wind Catcher/Tower natural ventilation system with performance testing in a real building. *International journal of ventilation,* 4 (4), 393-404.

[7] Chung, T. J. (2002). *Computational fluid dynamics.* New York, Cambridge.

[8] Jiang, Y., Allocca, C. And Chen, Q. (2004). Validation of CFD simulations for natural ventilation. *International journal of ventilation,* 2 (4), 359-370.

[9] Hughes, B. R. And Abdul Ghani, S. A. A. A.(2009). A Numerical Investigation Into The Effect of Windvent Louvre External Angle On Passive Stack Ventilation Performance. *Building and Environment*, 45 (4) 1025-1036.

In: Ventilation: Types, Standards and Problems
Editors: V. A. Romano and A. S. Duval, pp. 199-222
ISBN 978-1-61324-424-1

Chapter 8

VENTILATION SYSTEMS

M. I. El-Hadary* and A. E. Kabeel†
Tanta University, Faculty of Engineering,
Mechanical Power Department, Egypt

ABSTRACT

Ventilation is the replacement of stale air in a closed space with fresh air. It is a vital requirement for the comfort and health of closed space occupants. The reasons why ventilation is so important are to supply oxygen for breathing (respiration), to dilute pollutants such as body odours, to exhaled carbon dioxide, to remove unwanted heat from the closed space, to supply air for combustion appliances, to lower the relative humidity and so avoid condensation and to clear smoke in the event of fire.

Ventilation systems may be classified according their ability to supply and withdraw air from the closed space ventilated. It is common to different between the short cut principle, mixed principle, displacement principle and piston principle. Also, there are essentially some ways of ventilating any closed space as ventilation by displacement, ventilation by diffusion, natural ventilation, mechanical or forced ventilation, extract ventilation, supply ventilation and balanced ventilation.

1. INTRODUCTION

Ventilation is the replacement of stale air in a closed space with fresh air. It is a vital requirement for the comfort and health of closed space occupants.

Ventilation air, as defined in ASHRAE Standard 62.1[2] and the ASHRAE Handbook,[3] is that air used for providing acceptable indoor air quality. It mustn't be confused with vents or flues; which mean the exhausts of clothes dryers, and combustion equipment such as water , boilers, fireplaces, and wood stoves. The vents or flues carry the products of combustion which have to be expelled from the building in a way which does not cause harm to the

* E-mail: elhadary20032002@yahoo.com.
† E-mail: Kabeel6@hotmail.com.

occupants of the building. Movement of air between indoor spaces, and not the outside, is called transfer air.

Purposes of Ventilation

The purposes of ventilation are:

1. To provide a continuous supply of oxygen necessary for human existence.
2. To remove the products of respiration and occupation, that is; heat, moisture and carbon dioxide from people.
   - At rest a normal adult inhales between 0.10 and 0.12 L/s of air.
   - The exhaled breath contains between 3% and 4% of carbon dioxide, which is equal to 0.003 to 0.005 L/s.
   - The amount of heat from occupants is about 100 Watts sensible and 40 watts latent heat from a sedentary worker.
   - The amount of moisture produced by a sedentary person is about 59g of water vapour per hour.
3. To remove contaminants such as:
   - Water vapour
   - Heat and smells from cooking
   - Gases and vapours from industrial processes.
   - Formaldehyde from; insulation foam, furnishings, wallpaper, carpets, resin in wood products and plasterboard.
   - Outdoor aerosol pollutants such as; smoke, soot, mist, fumes, pollen, plant fibres, mould spores, viruses and bacteria
   - Indoor aerosol pollutants such as; carpet fibres, furniture fibres, clothing fibres, skin flakes, mites, viruses and bacteria.

In practice most ventilation systems for rooms inhabited by people dilute contaminants such as Heat and Carbon Dioxide to acceptable limits for short and medium term exposure.

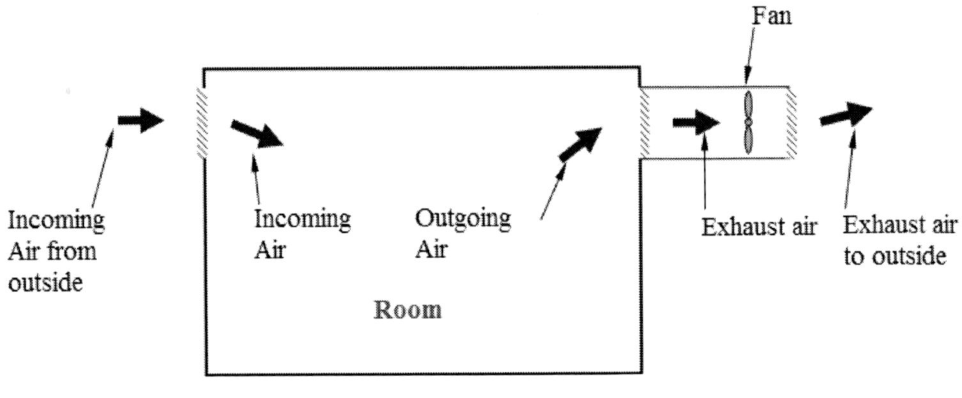

Figure 1. Typical Ventilation System [1].

Typical Ventilation System

A typical plant layout is shown in figure (1).

2. Ventilation Rates

The building regulations require those habitable rooms and toilets are to be vented by natural or mechanical means.

Table 1.

Building / Room	Air Change Rates - n - (1/hr)	Building / Room	Air Change Rates - n - (1/hr)
All spaces in general	min 4	Churches	8 - 15
Attic spaces for cooling	12 - 15	Club rooms	12
Auditoriums	8 - 15	Clubhouses	20 - 30
Bakeries	20	Cocktail Lounges	20 - 30
Banks	4 - 10	Computer Rooms	15 - 20
Barber Shops	6 - 10	Court Houses	4 - 10
Bars	20 - 30	Dental Centers	8 - 12
Beauty Shops	6 - 10	Department Stores	6 - 10
Boiler rooms	15 - 20	Dining Halls	12 -15
Bowling Alleys	10 - 15	Dining rooms (restaurants)	12
Cafeterias	12 - 15		
Dress Shops	6 - 10	Mills, textile dye houses	15 - 20
Drug Shops	6 - 10	Municipal Buildings	4 - 10
Engine rooms	4 - 6	Museums	12 -15
Factory buildings, ordinary	2 - 4	Offices, public	3
Factory buildings, fumes and moisture	10 - 15	Offices, private	4
Fire Stations	4 - 10	Police Stations	4 - 10
Foundries	15 - 20	Post Offices	4 - 10
Galvanizing plants	20 - 30	Precision Manufacturing	10 - 50
Garages repair	20 - 30	Pump rooms	5
Garages storage	4 - 6	Restaurants	8 - 12
Homes, night cooling	10 - 18	Retail	6 - 10
Jewelry shops	6 - 10	School Classrooms	4 - 12
Kitchens	15 - 60	Shoe Shops	6 - 10
Laundries	10 - 15	Shopping Centers	6 - 10
Libraries, public	4	Shops, machine	5
Lunch Rooms	12 -15	Shops, paint	15 - 20
Luncheonettes	12 -15	Shops, woodworking	5
Nightclubs	20 - 30	Substation, electric	5 - 10
Malls	6 - 10	Supermarkets	4 - 10
Medical Centers	8 - 12	Town Halls	4 - 10
Medical Clinics	8 - 12	Taverns	20 - 30
Medical Offices	8 - 12	Theaters	8 - 15
Mills, paper	15 - 20	Turbine rooms, electric	5 - 10
Mills, textile general buildings	4	Warehouses	2
		Waiting rooms, public	4

Table 2.

Activity	Minimum ventilation requirement (L /s per person)	
	0.5% CO2 limit	0.25% CO2 limit
Seated quietly	0.8	1.8
Light work	1.3 – 2.6	2.8 – 5.6
Moderate work	2.6 – 3.9	N/A
Heavy work	3.9 – 5.3	N/A
Very heavy work	5.3 – 6.4	N/A

Table 3.

Level of Smoking	Outdoor air supply rate (L/s per person)
No smoking	8
Some smoking	16
Heavy smoking	24
Very heavy smoking	36

2.1. Design Criteria

To design a ventilation system, the engineer has to meet two basic requirements:

1. To supply fresh air for the occupants.
2. To change the air in the room sufficiently so that smells, fumes and contaminants are removed.
   - The volume of fresh air (make up air) required for proper ventilation is determined of the size and use of the space.
   - The table (1): indicates recommended air change rates (air changes per hour) in some common types of rooms and buildings [2]:

The Table (2): gives Ventilation rates required to limit CO_2 concentration where level of activity is known [1]. Table (3): gives Recommended outdoor air supply rates for sedentary occupants [1].

3. Ventilation Systems

There are essentially some ways of ventilating a closed space [3]:

- Ventilation by displacement.
- Ventilation by diffusion.
- Natural ventilation.
- Mechanical or forced ventilation:
  1. Extract ventilation.
  2. Supply ventilation.
  3. Balanced ventilation.
- Spot cooling.

Ventilation by displacement, It is chiefly used to ventilate large industrial premises, as it can remove large volumes of impurities and heat if properly dimensioned. The air is supplied at low velocity directly into the occupied zone. This method provides excellent air quality, but is less suitable for offices and other smaller premises because the directional supply-air terminal device takes up a considerable space and it is often difficult to avoid some amount of draught in occupied areas.

Ventilation by diffusion, It is the preferable method for supplying air in situations requiring what is known as comfort ventilation. This is based on the principle of supplying air outside the occupied zone which then circulates the air in the entire closed space. The ventilation system must be dimensioned so that the air which circulates in the occupied zone is comfortable enough, in other words the velocity must not be too high and the temperature must be more or less the same throughout the zone.

An air stream which is injected into a closed space will attract, and mix together with, large volumes of ambient air. As a result, the air stream's volume increases while at the same time the air velocity is reduced the further into the closed space it travels. The mixing of the surrounding air into the air stream is termed 'induction'.

Natural ventilation, It is the flow of air through open windows, doors, grilles, and other planned closed space envelope penetrations, and it is driven by natural and/or artificially produced pressure differentials. A natural-draft roof ventilator should be positioned so that it receives the full, unrestricted wind. Turbulence created by surrounding obstructions, including higher adjacent closed spaces, impairs a ventilator's ejector action. The ventilator inlet should be conical or bell mounted to give a high flow coefficient. The opening area at the inlet should be increased if screens, grilles, or other structural members cause flow resistance. Closed space air inlets at lower levels should be larger than the combined throat areas of all roof ventilators. Stacks or vertical flues should be located where wind can act on them from any direction. Without wind, stack effect alone removes air from the closed space with the inlets.

Natural ventilation is ventilation without the assistance of fans or other mechanical air moving equipment. Natural ventilation uses no energy or little energy therefore reduces building running costs.

In large buildings where large amounts of air need to be changed in rooms, natural ventilation can be used if special features are used. Also is commercial building the ventilation system should be controlled to meet comfort criteria in rooms. This can be achieved by careful building design and using technology to provide adequate ventilation.

The diagram is shown in figure (2), shows a building layout that uses the stack effect to increase natural ventilation.

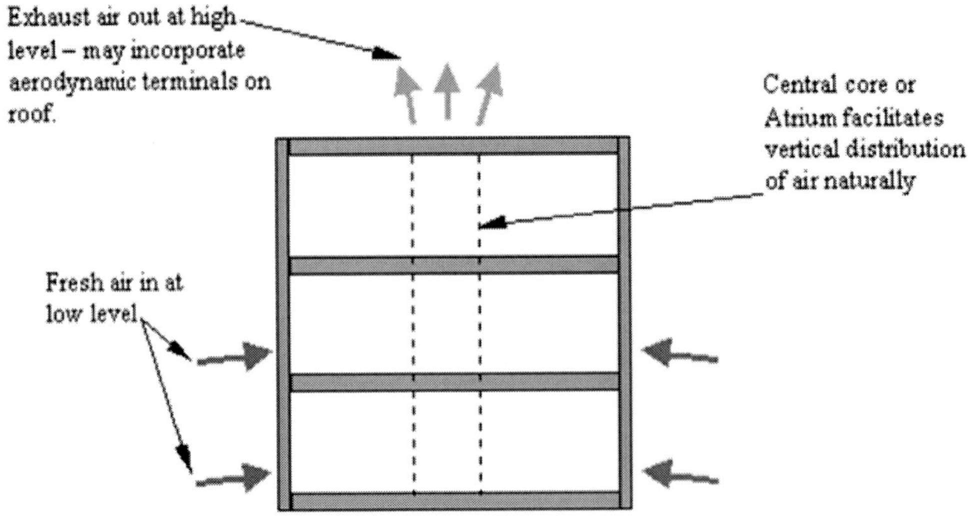

Figure 2. Natural Ventilation System [1].

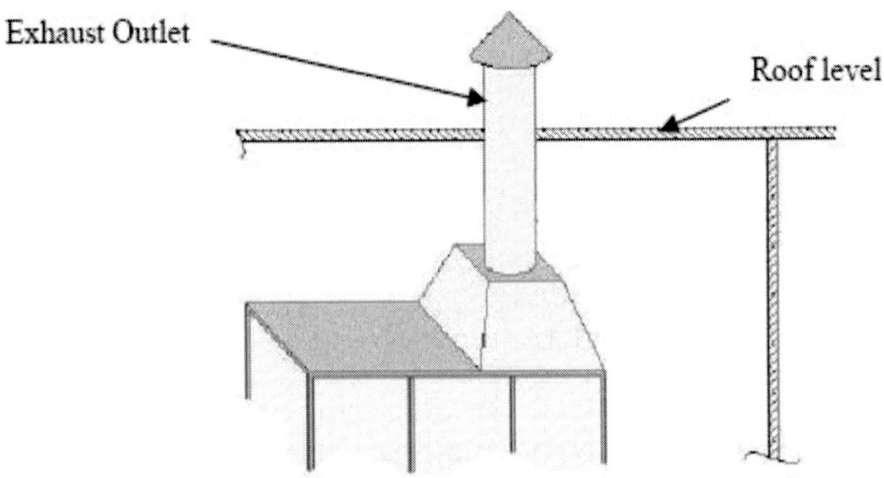

Figure 3. Exhaust outlets at roof level [1].

Exhaust outlets at roof level can be disguised or used as a feature of building design as shown in figure (3).

Another natural ventilation system uses the sun to assist air movement.

The vertical shafts in the building are glass fronted so that the sun heats up the air inside and causes it to rise out the openings at roof level.

The high level openings in this case are stainless steel chimneys.

As air flows out of the chimneys at roof level replacement air is drawn from the rooms into the shaft and thus naturally ventilated.

One of the difficulties with Natural Ventilation is that buildings have to be specially designed to incorporate features to assist air flow; these features may not look very good from the inside and outside and are sometimes disguised.

Effectiveness of Natural Ventilation

The effectiveness of natural ventilation for commercial buildings depends on several criteria.

These are wind strength and direction, size of openings, air temperatures and height of building.

For effective controlled ventilation the designer should not rely solely on the wind but more on the stack effect and air controls.

Dampers can be used to control air entering and/or exiting a natural ventilation system.

These dampers could be linked to occupancy sensors, temperature sensors, time switches and other weather sensors to give automatic control of ventilation which is the key to a useful system.

The diagram below shows some of the features of a Natural Ventilation system for a four-storey building.

Stack Driven Ventilation Calculations

Stack ventilation calculations in the simplest form ignore wind effects, although these can be allowed for in a more complex analysis.

The pressures developed in stack systems can be determined from the equation (1) [1].

$$\delta P_s = - \rho_{ins} . g. T_{ins} (h_2 - h_1) (1/T_{out} - 1/T_{ins}) \qquad (1)$$

where;

δP_s = Stack Effect Pressure (Pa)
ρ_{ins} = Air density inside stack (kg/m^3)
g = Acceleration due to gravity (9.81 m/s^2)
h_1 = Height of intel of stack above datum (m)
h_2 = Height of outlet of stack above datum (m)
T_{out} = Temperature of air outside stack (^oK)
T_{ins} = Temperature of air inside stack (^oK)

The equation (2) can be used to determine air flow rates in stack driven ventilation or the opening areas required [1].

$$Q = C_d . A [(2 / \rho_{ins}) \ \rho_{ins} . g. (h_{npl} - h) (T_{ins} - T_{out} / T_{ins})] \qquad (2)$$

where;

Q = Air flow rate through a large opening (m^3/s)
C_d = Discharge coefficient (0.61 for large openings)
A = Opening area (m^2)
ρ_{ins} = Air density inside stack (kg/m^3)
g = Acceleration due to gravity (9.81 m/s^2)

h_{npl} = Height of neutral pressure level above datum (m)
h = Height of opening above datum (m)
T_{out} = Temperature of air outside stack (^oK)
T_{ins} = Temperature of air inside stack (^oK)

Mechanical (forced) ventilation, It is the intentional movement of air into and out of a closed space using fans and intake and exhaust vents; it is also called mechanical ventilation.

Forced ventilation can be used to supplement both horizontal and vertical ventilation. Forced ventilation is the process of mechanically moving air inside a structure. The three primary methods of forced ventilation are hydraulic, negative, and positive pressure. It is the goal of this article to begin the discussion of forced ventilation by examining the different tools, techniques, advantages, and disadvantages of hydraulic and negative pressure ventilation.

Some mechanical ventilation systems use fans for both supplying and extracting air, thus mechanical ventilation systems may be classified as follows:

1. Supply system
2. Extract system
3. Balanced system.

1. Supply Ventilation System

- Fresh air is supplied to a space from outside as shown in figure (4); this air provides oxygen for breathing and ventilation for occupants.
- Air is removed from the space by 'natural' means since the room is pressurised by the supply air.
- In some cases it is advantageous to heat incoming air to offset fabric losses and avoid cold draughts in winter.
- This is known as a plenum system and is useful in large rooms with a high ceiling, which can be difficult to heat with radiators.

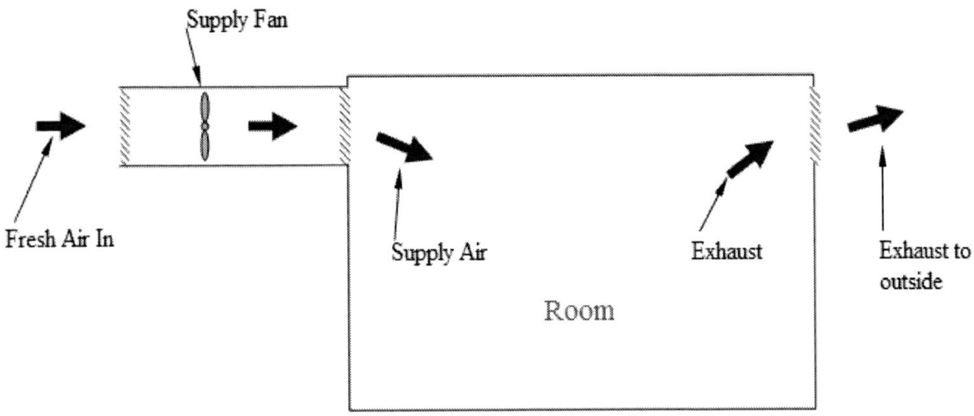

Figure 4. Supply Ventilation System [1].

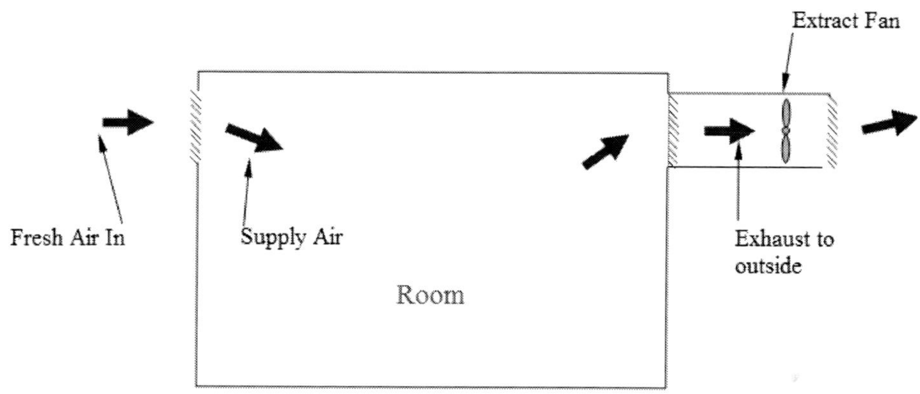

Figure 5. Extract Ventilation System [1].

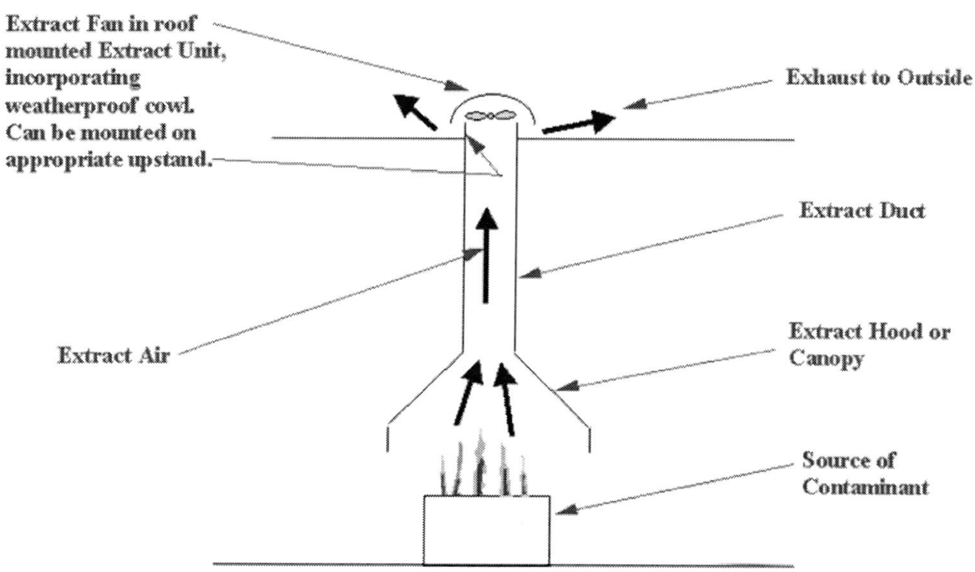

Figure 6. Simple extract system for Industrial Places [1].

1. Extract Ventilation System

- The principal function of an extract ventilation system is the removal of an unwanted contaminant, whether it is solid, gaseous or thermal.
- Air is extracted from the space and replaced by fresh air entering from outside; the space is under negative pressure; therefore air is naturally drawn into the building as shown in figure (5).
- In industrial ventilation, airborne dust, toxic fumes, vapours and excessive heat have to be removed.
- This is sometimes carried out at source, thus minimising the contamination of the occupied space.

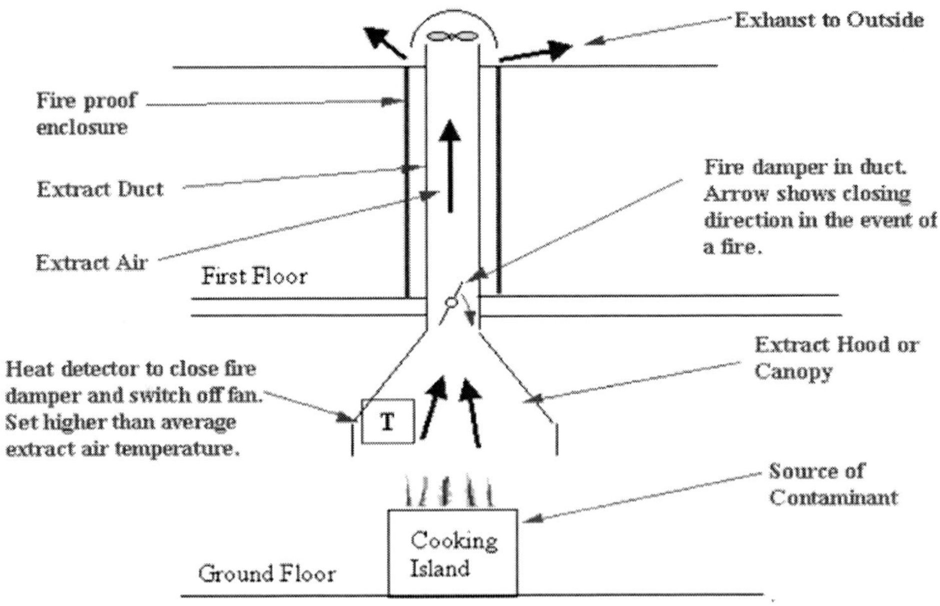

Figure 7. Simple extract system for a kitchen [1].

- The spread of contaminant is decreased by installing hoods or canopies over the source and connecting these to the extract ventilation system as shown in figure (6).
- In kitchen extract ventilation it may be necessary to install a fire damper and enclose the duct which passes through upper floors in a fire resisting shaft.
- This is because there is a possible danger of fire and smoke spreading throughout a building in the ductwork system.
- Obviously when designing for the event of a fire in the building one should be careful to provide a safe and protective ventilation installation.
- Consult all relevant standards and get as much advice as possible especially from the local Fire Authority.
- A simple extract system for a kitchen is shown in figure (7).

2. Balanced Ventilation System

A balanced system enables full control of ventilation to be achieved by the use of separate mechanical supply and extract systems.

It is usual to provide a surplus of supply air over extract air so as to maintain the pressure in the building at a slightly higher pressure than outside.

This minimises natural infiltration which reduces the likelihood of draughts.

A typical system is shown in figure (8a).

It is a good idea to filter outside air so that atmospheric pollutants are excluded.

Also in winter, cold outside air may cause discomfort so the fresh air is heated. This can be used to offset heat losses thus providing the means of heating the building or room as shown in figure (8b).

This is the same as a plenum heating system with return air system.

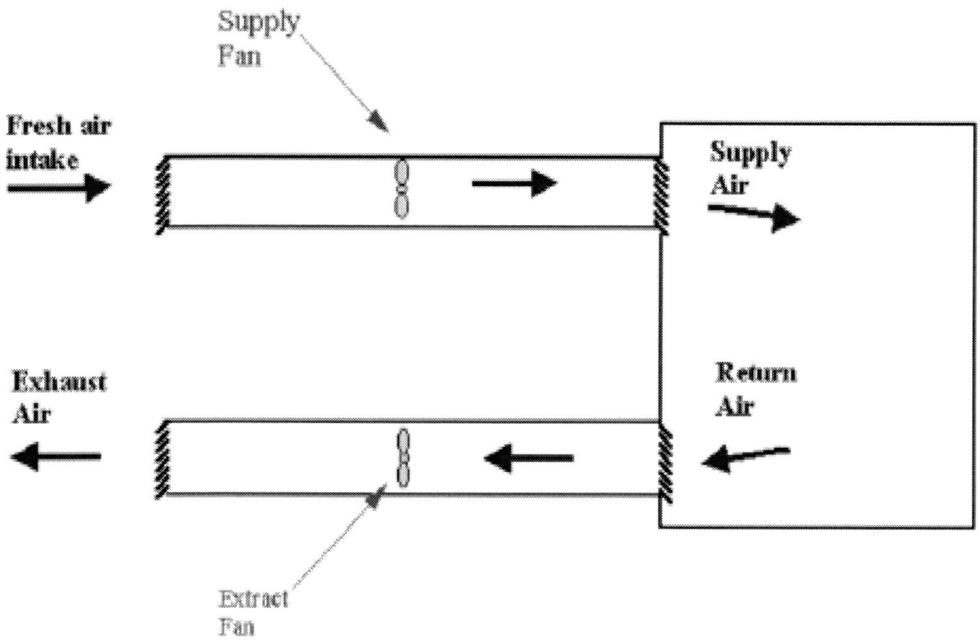

a.

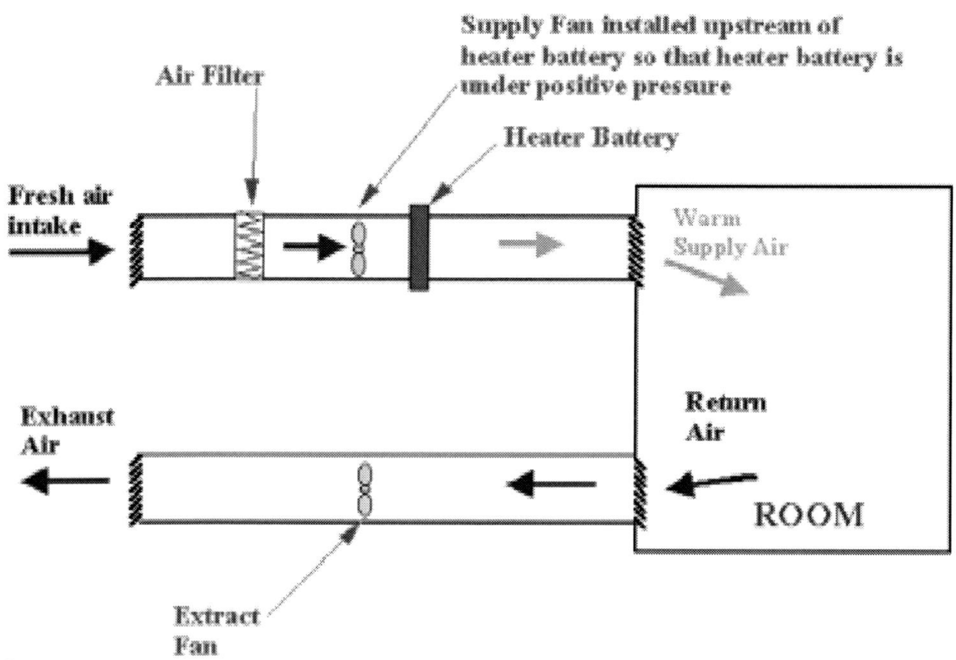

b.

Figure 8. (Continued).

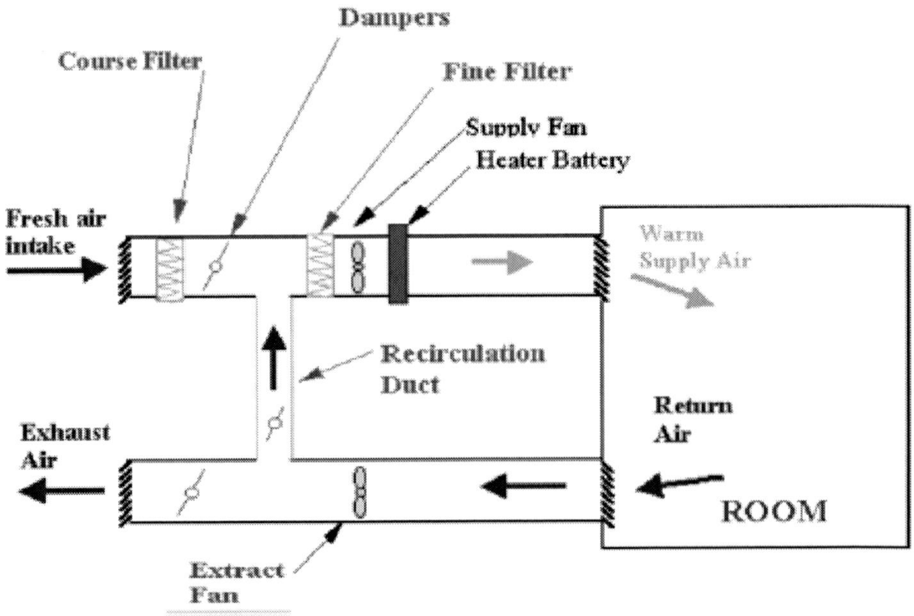

c.

Figure 8. Balanced Ventilation Systems [1].

In most balanced systems, the supply air quantity, which is required, works out to be much more than that needed for fresh air supply to occupants.

It is possible therefore, to recirculation some of the extract air back into the supply duct to make use of the heat which it contains as shown in figure (8c).

The amounts of fresh air in each section of ductwork are controlled by dampers, which can be set during commissioning so that the design quantity of air with the correct proportion of fresh air is supplied to the space.

Recirculation only works if the air has not been contaminated in the space. In kitchens, toilets, smoke filled spaces, etc., where the air contains odours or other contaminants all the extract air is removed and no recirculation takes place.

Filters are usually fitted in supply and balanced ventilation systems to remove any airborne particles in the fresh air intake duct.

A finer filter may be installed in a balanced ventilation system after the mix point to remove dust generated within the space.

The bag filters are for collecting fine particles of dust and are sometimes referred to as fine filters.

Spot cooling [3], Energy costs continue to climb, at the same time, utilities are frequently unable to provide all of the electricity that industry needs. When these circumstances are combined with the urgent need to control capital expenses, the case for spot cooling becomes very strong.

For many industries and companies, in a wide variety of applications, spot coolers are the answer to controlling both energy use and capital expenses.

Quite simply, spot cooling is a method of cooling overheated areas within a larger area, as opposed to providing general cooling by means of a remote centrally controlled air conditioning system. In operation, spot cooling is an extremely efficient means of cooling people, processes, and equipment because it directs a localized stream of cool air exactly where it is needed.

The concept of spot cooling is not a new idea, but the use of spot coolers to accomplish localized cooling within work environments and equipment closed spaces is a recent development.

If the workplace is located near a source of radiant heat that cannot be entirely controlled by radiation shielding, spot cooling is recommended. The air temperature of the workplace is limited to 20° to 30°C, depending on the heat load and the air speed.

Common cooling systems include refrigeration equipment, evaporative coolers, and high-velocity ventilating air. For manufacturing operations, particularly heavy industry where mechanical cooling cannot be economically justified, evaporative cooling systems often provide good working conditions. If the operation requires heavy physical work, spot cooling, evaporative cooling, or refrigerated air can be used. To minimize summer discomfort, high outdoor air ventilation rates may be adequate in some hot process areas. In all these operations, a mechanical air supply with good distribution is needed.

In hot workplaces that have only a few work areas, it is impractical and wasteful to maintain a comfortable environment in the entire closed space. However, working conditions in areas occupied by workers can be improved by air-conditioned cabins, individual cooling, and spot cooling.

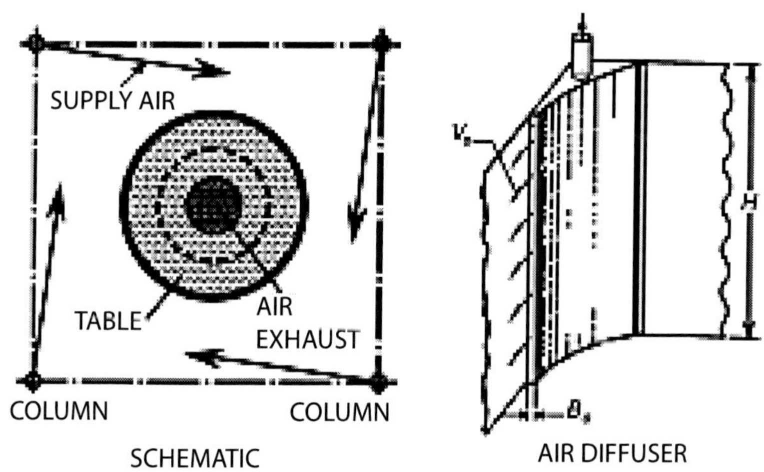

Figure 9. Spiral Vortex Ventilation System [1].

Air-conditioned cabins provide thermal comfort effectively, but they are expensive. Control closed spaces for monitoring production and manufacturing processes can use this technology.

The supply air can be cooled by a conventional heat exchanger or by a vortex tube explained in figure (9), the suits are simple and self-regulating, provided that there is sufficient airflow.

When the supply air is at skin temperature, heat is removed by increasing the evaporation of sweat. Lowering the supply air temperature improves the convection heat loss, thereby increasing the cooling capacity. Individual cooling with vests or helmets is appropriate if the thermal conditions are not extreme. Vests and helmets remove less heat than suits, but they cost less, are easier to use, and allow increased mobility.

Spot cooling is probably the most popular method of improving the thermal environment. Spot cooling can be provided by radiation (decreasing the mean radiant temperature), by convection (increasing the air velocity), or by a combination of the two methods. Spot cooling equipment is fixed at the workstation, whereas individual cooling has the worker wearing the cooling equipment.

Spot cooling is useful in closed spaces with elevated temperatures, regardless of climatic or geographical location. When the dry-bulb temperature of the air is below skin temperature, convection rather than evaporation cools workers. A 27°C air stream feels so good that its relative humidity is inconsequential.

Spot cooling permits the driving of long headings for exploration or extended development prior to the installation of primary ventilation equipment. Using mechanical refrigeration systems with direct-expansion air coolers spot cooling services a single heading or localized mining area. Spot coolers allow development headings to be advanced more rapidly and under more comfortable conditions.

4. Ventilation System Design

Design Methodology

When considering ventilation system design the following approach could be adopted before sizing begins and the following questions should be considered:

1. What areas need ventilation?
   The contaminants should be listed for these areas.
2. What type of system should be used, supply, extract or balanced?
3. Are there any alternative systems to consider?
4. Is air conditioning necessary in the building?
   If air conditioning is necessary then should it be incorporated into the ventilation system?
5. Where should the fan(s) and plant be installed?
6. What type of fan(s) and plant should be used?
7. Is a separate heating system necessary?
8. What type of control system should be used?
9. What type of air distribution system should be used, upward or downward?
10. Have you considered what will happen in the event of a fire in the building?
11. Have you considered the noise from fans?

After all the above questions have been answered the sizing process may commence.

Sizing

The sizing procedure is as follows:

1. Calculate Ventilation rates.
2. Decide on number of fans and grilles/diffusers.
3. Draw scale layout drawing:
   - Position fan(s).
   - Lay out ductwork.
   - Lay out grilles and diffusers.
   - Indicate flow rates on drawing.
4. Size ductwork
5. Size fan
6. Size grilles and diffusers.

Good Ventilation Design

Good ventilation design should have the following features:

1. Not noisy
2. Concealed
3. No draughts
4. Efficient fan
5. Good control of air flow with dampers and appropriate diffusers.
6. Good control of room temperature.
7. Appropriate duct sizes.
8. Well supported ducts and equipment.
9. Prevent spread of smoke in the event of a fire with smoke/fire dampers.
10. Ensure that supply air is clean by using a filter.
11. Ensure that vermin cannot enter the duct system by using a bird/insect screen in the fresh air intake.
12. Minimise risk of infection in some buildings (e.g. hospital) by having no recirculation duct.
13. Use recirculation duct in some buildings to save energy.
14. Use appropriate air change rates to meet room requirements.
15. Use appropriate fresh air rate to meet room occupants' requirements.
16. Use suitable system to fit in with building aesthetics.
17. Avoid duct leaks by using proper jointing method.

Drawings for Ventilation Systems

A good drawing should show as much detail as is necessary and be clear and easy to read. It is a good idea to remember what the drawing is for or who is going to read it to obtain information.

The drawing may be used by;

- An installer or ventilation contractor.
- A draughtsman /CAD designer to produce a more detailed drawing.
- A main contractor or builder.
- A sub-contractor – electrical, controls, heating/plumbing.
- An estimator or quantity surveyor.
- A clerk of works.
- A commissioning engineer.

The following information and details should be produced in a ventilation drawing;

Drawing Information

1. Draw outline of building.
2. Show cavity wall
3. Show internal partitions, doors and windows.
4. Draw diffusers / grilles.
5. Show kitchen canopy.
6. Show ducts as 2 lines.
7. Show branches as; shoes / boots / pops.
8. Show direction of air flows.
9. Add notes for; Duct sizes and air flow rates.
10. Position fans
11. Add note for fan size.
12. Add Index run identification i.e. sections.
13. Add Title block showing;
    - Building Services Consultants name.
    - Job, job address.
    - Client
    - Drawing description
    - Drawn by
    - Date
    - Drawing Number
    - Scale
    - Revisions.

Further Information to be Given

1. Show other building details such as; stairs, lifts, structure, roof, roofspace, ceiling spaces, void spaces, shafts.
2. Show steelwork details where appropriate.
3. Show flexible ductwork.
4. Show grid pattern for ceiling tiles and building grid if appropriate.
5. Show table of diffusers and grilles.
6. Show table of External louvers.
7. Show table of Fans with duty and specification
8. Show furniture.
9. Show Volume Control Dampers (VCD's)
10. Show fire dampers.
11. Show Attenuators if appropriate.
12. Show what symbols represent in a legend e.g. FD – Fire damper.
13. Show any other equipment such as; AHU (Air Handling unit), Chiller, cassette unit, air cooled condenser, heater battery, cooling coil, fan coil unit, etc.
14. Installation Notes and fixing methods.
15. References to other drawings e.g. steelwork drawing.
16. Drawing details of important areas e.g. fixing, weatherproofing.

Some of the above details may be given in a specification but it is a good idea to produce the same information again on a drawing if space permits to give a full picture of the job.

One of the reasons for this is that the drawings or drawings that are used for installation purposes often don't tell the whole story and mistakes can be made.

Also installers don't always have access to specification documents.

One must remember that if changes are made to equipment then both specification and drawings should be updated.

5. Ventilation Calculations

The following formulae may be used:

A. For General Mechanical Ventilation

Ventilation rate (m^3/h) = Air Change Rate (/h) x Room Volume (m^3) (3)

Air Change Rate expressed in SI-units:

Ventilation rate (m3/s) = Ventilation rate (m3/h) / 3600 (4)

B. For Calculating Fresh Air Ventilation Rates

Fresh Air Rate (m3/s) = Fresh Air rate per person (l/s/p) x Number of occupants (5)

Fresh Air rate per person (L/s/p) comes from Table (3).

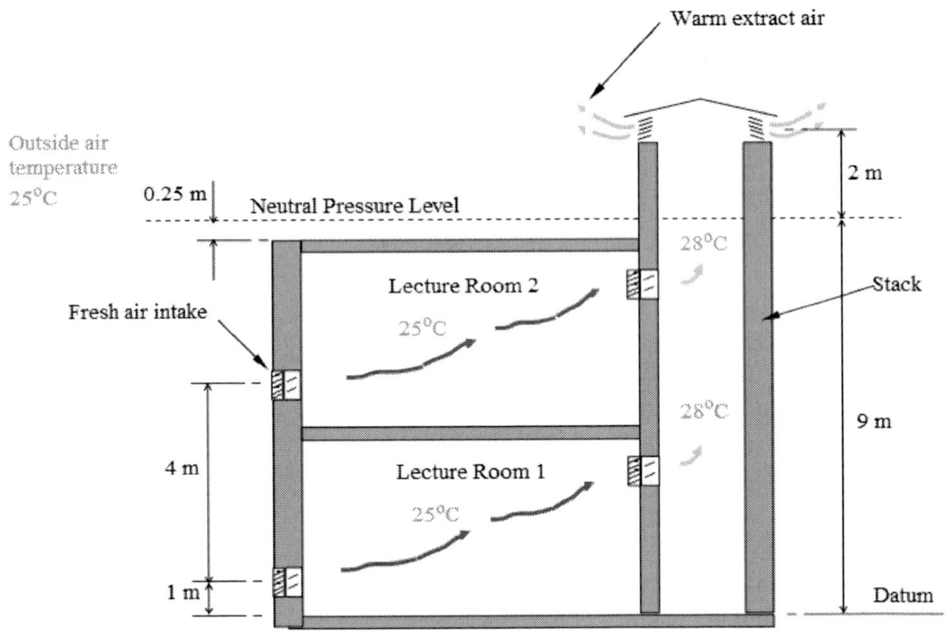

Figure 10. Example (2) [1].

6. Examples

Example (1)

With an air flow of 3 m^3/s in a 20000 m^3 room the air flow rate can be calculated as n = 3600 (3 m^3/s) / (20000 m^3) = 0.54

Example (2)

Calculate the ventilation opening area required in a Stack ventilation system for the building shown in figure (10).

Data

The flow rate required each room is 4 air changes per hour.
Each lecture room measures internally 24 m x 10 m x 4 m high.

Answer

Air flow rate for each room Q= Room volume x Air change rate / 3600

$Q = (24 \times 10 \times 4) \times 4 / 3600$
$Q = 960 \times 4 / 3600 = 1.07\ m^3/s$

Rearranging above formula for Area (A) gives;

$$A = Q / C_d\ [(2 / \rho_{ins})\ \rho_{ins} \cdot g \cdot (h_{npl} - h)\ (T_{ins} - T_{out} / T_{ins})]$$

For Ground floor room;

$A = 1.07 / 0.61\ [(2 / 1.1656)\ 1.1656 \times 9.81\ (9 - 1)\ (301 - 298 / 301)$
$A = 1.07 / 0.61 \times 1.565$
$A = 1.121\ m^2.$

For First floor room;

$A = 1.07 / 0.61\ [(2 / 1.1656)\ 1.1656 \times 9.81\ (9 - 5)\ (301 - 298 / 301)$
$A = 1.07 / 0.61 \times 0.782$
$A = 2.242\ m^2.$

Note: The upper floor has less stack suction pressure so openings are larger.

Example (3)

Design a ventilation system for the building shown in figure (11).
The building is a small workshop used for general metal fabrication and welding.
No welding booths are installed.
If the building gets too warm the door may be opened for free cooling.

Data

Building dimensions: 15.0 m long x 7.0 m wide x 5.0 m high to eaves.
Roof ridge height is 7.5 m.
The air change rate for mechanical ventilation is to be 6 air changes per hour.
The building volume may be calculated from:

$V = (L \times W \times H_{to\ eaves}) + (0.5 \times W \times \text{perpendicular height from eaves to ridge} \times L)$
$V = (15 \times 7 \times 5) + (0.5 \times 7 \times 2.5 \times 15)$
$V = 525 + 131.25 = 656.25\ m3.$

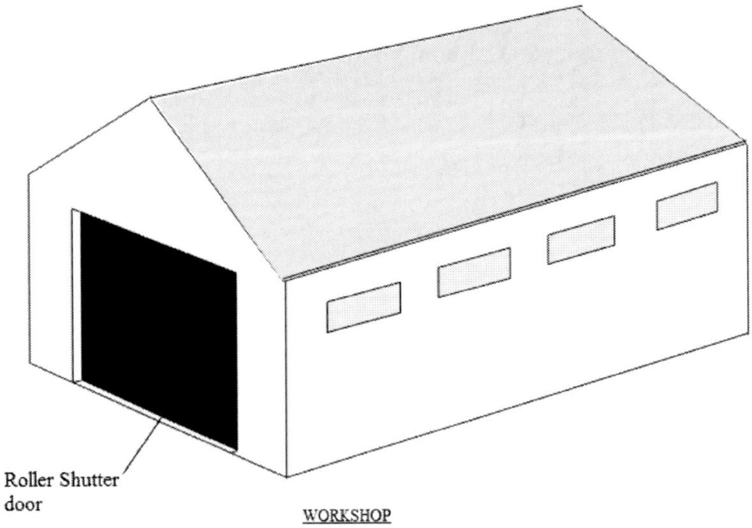

Figure 11. Example (3) [1].

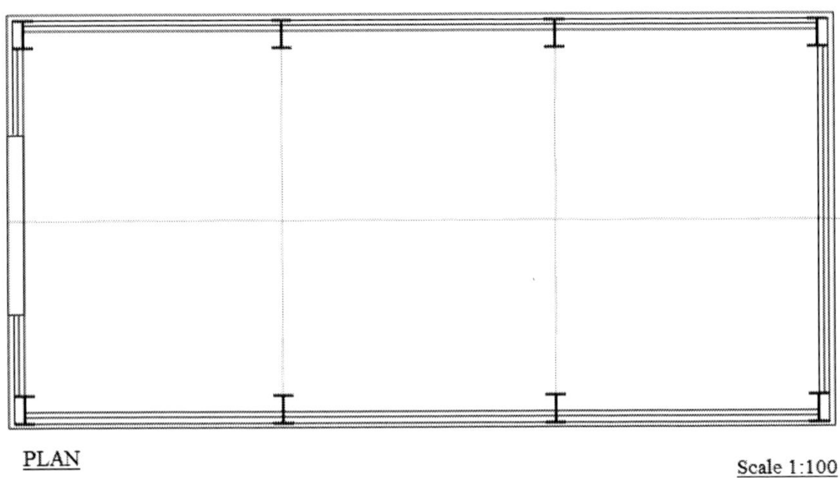

Design Methodology

1. What areas need ventilation?
   Answer: The total area of the workshop.
   The contaminants are welding fumes and heat.
2. What type of system should be used, supply, extract or balanced?
   Answer: Extract.
3. Are there any alternative systems to consider?
   Answer: No, supply air is not required.
4. Is air conditioning necessary in the building?
   Answer: No, external door to be opened.
5. Where should the fan(s) and plant be installed?
   Answer: On the roof.

6. What type of fan(s) and plant should be used?
   Answer: Roof mounted units with weatherproof cowl.
   Anti-backdraught shutters on each fan. No ductwork.
7. Is a separate heating system necessary?
   Answer: No, not for extract only.
8. What type of control system should be used?
   Answer: Variable speed fan(s) for a flexible system.
9. What type of air distribution system should be used, upward or downward?
   Answer: Upward for extract systems, with duct mounted extract grilles.
10. Have I considered what will happen in the event of a fire in the building?
    Answer: The fans could remove smoke in the event of a fire.
11. Have I considered the noise from fans?
    Answer: In a workshop, fan noise is not a problem.

Sizing

The sizing procedure is as follows:

1. Calculate Ventilation rates.
2. Decide on number of fans and grilles/diffusers.
3. Draw scale layout drawing:
   - Position fan(s).
   - Layout ductwork.
   - Layout grilles and diffusers.
   - Indicate flow rates on drawing.

1. Ventilation Rates

For Extract ventilation systems the rate in air changes per hour is obtained for some buildings in Table (1).

There is no explicit rate given for Factories and warehouses or Industrial ventilation due to the variations in building design and usage.

The rate for this example is given in the question as 6 air changes per hour.

The following formulae may be used:

For General Mechanical Ventilation

Ventilation rate (m^3/h)	=Air Change Rate (/h) x Room Volume (m^3)
	= 6 x 656.25
	= 3937.5 m^3/h
Ventilation rate (m^3/s)	= Ventilation rate (m^3/h) / 3600
	= 3937.5 m^3/h / 3600
	= 1.094 m^3/s.

Three extract roof units are used so each will remove 1.094 / 3 = 0.365 m^3/s.

The resistance to air flow is negligible since there is no ductwork or grilles connected to the fans.

It is a good idea to look at several types of extract unit to ascertain the best option for the job. A typical unit is shown below.

Choosing a Fan

To choose a suitable fan one must look at the performance curves.

These curves show the pressure developed by a fan at a given flow rate.

The operating point of the system is marked as a point on the curve. To accurately establish a fan size the air flow resistance would have to be calculated.

Figure 12. Roof Extract Fan [1].

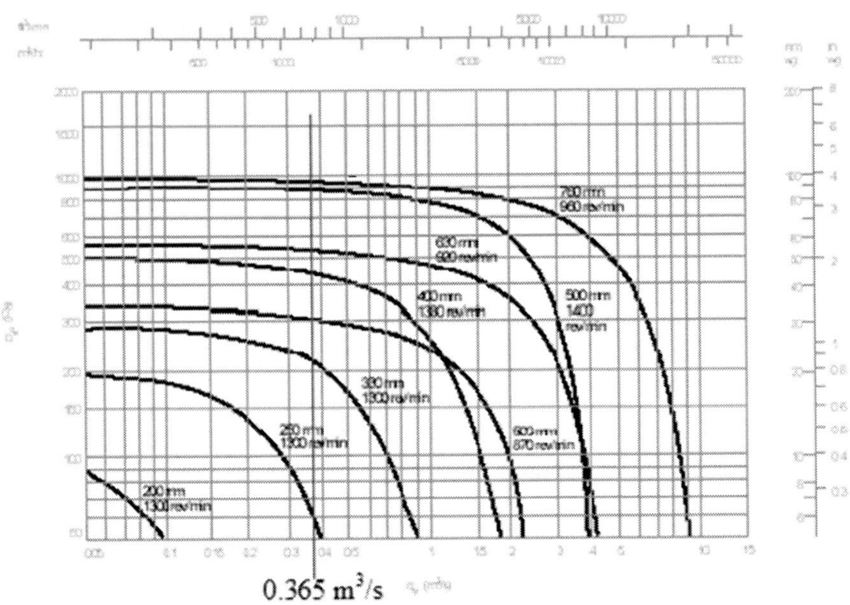

Figure 13. Fan Performance Curve [1].

REF.	FAN TYPE	SIZE Dia. (mm)	VOLUME FLOW RATE (m^3/s)	PRESSURE (Pa)	SPEED (rpm)	SOUND LEVEL (dB)
Woods DSP 315	Propeller	315	0.37	32	1420	48
Woods DSM 330	Mixed flow	330	0.37	230	1300	57
Woods DSC 250	Centrifugal	250	0.37	135	1350	51
Woods DSJ 315	Axial flow	315	0.37	55	1420	49

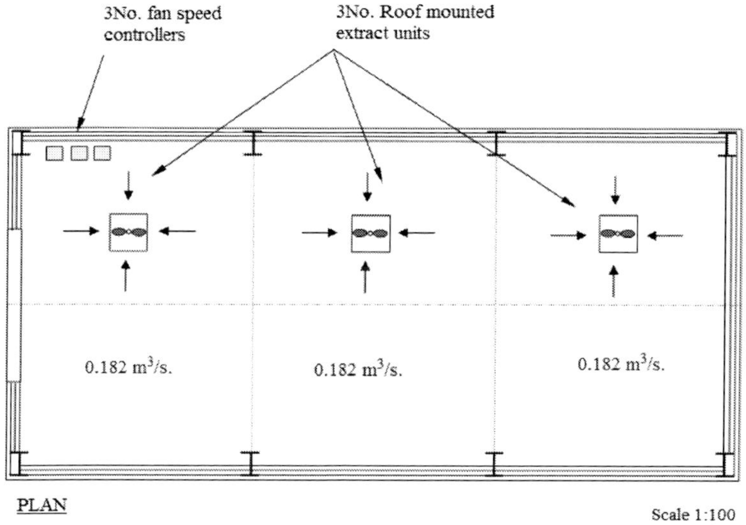

Figure 14. layout drawing [1].

In the absence of duct resistance data we can look at typical fans for the job. There are several options to choose from, for example, from the Woods Air Movement Ltd catalogue:

From the table (4): the propeller type fan and the axial flow fan produce 32Pa and 55 Pa pressure which is sufficient for the proposed system.

The noise output from both fans is similar.

I would choose the DSP 315 propeller type unit since a speed controller is available.

Conclusions

Finally, this chapter explains that the ventilation system is very important in the design of any residential or industrial building.

Acknowledgments

Thanks go first to ALLAH. I wish to express my appreciations and deep gratitude to my supervisors Prof. Dr. Gamal Sultan and Prof. Dr. Abd-Elnaby kabeel and Eng. Alaa Abo-freikha for their continuous guidance and encouragement throughout the course of this work.

I wish also to thanks my father, mother and wife who gave me more than support. They gave me the power to continue this research. I think that there are no words that can express my love and grateful for them. I express my gratitude to my family because this which remains at the end in the heart.

REFERENCES

[1] http://www.arca53.dsl.pipex.com/.

[2] http://www.engineeringtoolbox.com/air-change-rate-room-d_867.html

[3] A.E. Kabeel, G. I. Sultan, Z. A. Zyada and M. I. El-Hadary," Performance Study of Spot Cooling of Cabinet" M.SC. Thesis, Chapter 1, 2008.

[4] Ventilation and Infiltration chapter, Fundamentals volume of the ASHRAE Handbook, ASHRAE, Inc., Atlanta, GA, 2005

[5] ANSI/ASHRAE Standard 62.1, Ventilation for Acceptable Indoor Air Quality, ASHRAE, Inc., Atlanta, GA, USA

[6] The ASHRAE Handbook, ASHRAE, Inc., Atlanta, GA, USA

[7] Kavanaugh, Steve. Infiltration and Ventilation In Residential Structures. February 2004.

[8] http://epb.lbl.gov/Publications/lbnl-54331.pdf

[9] Apte, Michael G. Associations between indoor CO2 concentrations and sick building syndrome symptoms in U.S. office buildings: an analysis of the 1994-1996 BASE study data." *Indoor Air*, Dec 2000: 246-258.

[10] http://www.wapa.gov/es/pubs/techbrf/co2.htm

[11] How Natural Ventilation Works by Steven J. Hoff and Jay D. Harmon. Ames, IA: Department of Agricultural and Biosystems Engineering, Iowa State University, November 1994.

[12] ASHRAE Handbook of Fundamentals, Chapter 26 by American Society of Heating, Refrigeration and Air-Conditioning Engineers (ASHRAE). Atlanta, GA: 2001.

[13] http://www.engineeringtoolbox.com/ventilation-efficiency-d_124.html

In: Ventilation: Types, Standards and Problems ISBN 978-1-61324-424-1
Editors: V. A. Romano and A. S. Duval, pp. 223-233

Chapter 9

ADVANCES IN NATURAL VENTILATION FOR ENERGY SAVING AND THERMAL COMFORT

José A. Orosa[1,*] and Rafael Santos[2,†]

[1]Department of Energy and Marine Propulsion, University of A Coruña, Campus de Riazor, A Coruña, España

[2]Department of Industrial Engineering, University of A Coruña, Campus de Serantes, Ferrol, España

ABSTRACT

Indoor air is a fundamental parameter to be considered at the building design moment. In this sense, an adequate selection of a ventilation method must be done among natural, mechanical and hybrid ventilations. In recent research works, based on the tracer gas decay method, these three ventilation methodologies were compared in urban environments to define their performances in accordance with the nominal time constant, which is the inverse value of air-exchange rates and air-exchange efficiency. These results showed that hybrid ventilation is associated with rather higher nominal time constants and higher air exchange efficiency values than natural cross-ventilation. It was obtained that ventilation efficiencies seem to be higher when the nominal time constants are higher or the air-exchange rates are lower.

In this chapter some different methods of natural ventilation and what the current research works state, step by step, are shown. In particular, these steps are classified on the basis of field measurements, laboratory experimental data and simulation tools. Finally, future ventilation research topics like energy saving, thermal comfort and personalized ventilation methods are presented. The general obtained conclusion is that residential mixing ventilation and industrial hybrid displacement ventilation are very interesting alternatives. Furthermore, this discusses how a good agreement between developed models and experimental results is obtained and, subsequently, it can be employed to predict the relative performances of ventilation systems.

---

* Paseo de Ronda 51, 15011. A Coruña, España.

† Campus de Serantes, Avda. 19 de Febrero S/N. 15405, Ferrol, España.

1. Problems in Indoor Ambiences

Currently when ventilation in residential and industrial buildings, and hospitals and classrooms is considered, ambience is a subject of special interest. With regard to residential ambience, it is the environment where occupants spend most of their daily time and, in consequence, so health and higher energy consumption issues are influenced by ambiences. On the other hand, industrial ambiences have higher possibilities to present work risk problems that at the same time depend on the kind of indoor activities developed. Hospital ambiences can result in infection risks among patients and, in consequence, the ventilation method must be defined considering the kind of diseases. Finally, schools and offices represent an interesting indoor ambience where energy saving in air renovation is of special interest but always within adequate CO_2 levels. Once we have defined the main parameters that control the indoor air ventilation, some current real field studies are summarized to define the actual state of indoor ambiences.

1.1.1. Indoor Ambiences in Residential Buildings

One of the most important local indoor ambiences that is being analyzed in the last years is that of bathrooms, because these are humidity and odor sources, along with the kitchen and human release, during the night. In consequence, the ventilation rate, based on an adequate building design, appears to be nearly the only one solution for these problems.

In this sense, to analyze current bathroom ventilation problems, recent research works [1, 2] have compared different ventilation schemes for removal of the unpleasant odor of bathrooms. To analyze their ventilation performance some particular indices like the age of the air, the air exchange efficiency and the odor removal efficiency were employed. The air exchange efficiency is the efficiency of flushing the bathroom with the outdoor air and the odor removal efficiency denotes the ventilation efficiency of the bathroom. All these parameters were defined by the tracer gas decay method and constant injection method.

The results showed that, as it was expected, the negative pressure difference within the bathroom prevents unpleasant odors from entering into the neighboring rooms. Finally, it was concluded that the ventilation arrangement of forced-ceiling supply and wall-exhaust systems present the best ventilation performance to extract odors out of the bathrooms, among the three studied ventilation schemes.

As it was commented before, bedrooms are sources of CO_2 and humidity and these must be controlled during the night by different ventilation methodologies. In particular, these parameters are of special interest as health hazards. For example, recently findoor environments in rooms of children with health problems like asthma and allergy were developed [3]. In this research works the methodology to analyse indoor ambiences in bedrooms was based into measure the concentration of carbon dioxide over an average of 2.5 days and the ventilation rate during the night was defined in accordance with the mass balance for the occupant generated CO2. rom these measures it was concluded that only 32% of the bedrooms had an average CO2 concentration below 1000 ppm during the night and 23% above 2000 ppm and 6% of the room experienced concentrations above 3000 ppm.

As a general conclusion we can say that not only the total air changes of a global indoor ambience but also the relative air changes between rooms must be considered. For example, the air change rate calculated from the generated CO_2 concentration was found to be between 0 and 51%, lower than the total air change rate and it was associated with the position of the room in relation to the adjacent rooms, occupancy in the adjacent rooms, the nominal air change rate and room-to-room airflows.

1.1.2. Ventilation in Industrial Ambiences

Industrial ambiences house different contamination sources that can lead to, under an inadequate ventilation rate, health risks; so to control these architects and workers should think of improving some industrial ventilation methods like mixing ventilation systems (MVSs). MVSs consist in an injection of cold air from the upper part of the room reaching a turbulent mixing of the cold stream with the ambient air that determines the cooling of the whole volume and establishes a uniform temperature level with a very low vertical temperature gradient.

In recent research works [4] an easily retrofitted improvement solution to existing MVSs, called retrofit hybrid displacement ventilation system (RHDVS), is suggested and compared with the traditional MVS taking into account the technical and economic aspects. In this research work, experimental results show that in three buildings housing several manufacturing lines that hybrid displacement ventilation is more effective in establishing a correct flow field in the area dominated by personnel operation. It also allows to obtain a significant vertical temperature gradient which makes it possible to establish comfortable environmental conditions at floor level without the need for conditioning the entire room volume that let us increase the cooling air temperature and, in consequence, the power consumption of the cooling plant.

As a general conclusion we can say that the upgrade to hybrid displacement ventilation is a very interesting alternative in industrial ambiences, at least to cool rooms with a high ceiling (high internal volumes) that makes mixing ventilation a poorly cost-effective solution.

1.1.3. Ventilation in Special Environments

Indoor ambiences in schools and offices indicate a need for high energy consumption during the year. In consequence, new heating ventilation and air conditioning (HVAC) techniques are developed and being improved in the last years. Some methods are centred on the own heating and cooling systems and recently, in the control system that governs the HVAC equipment. This last method is a low-cost solution to reduce the energy consumption with an existing HVAC system and, in consequence, it is the first indoor maintenance step to be analyzed.

Some recent research works tried to improve the HVAC control system and define how much Demand Control combined with Displacement Ventilation (DCDV) reduce the ventilation air volumes and the energy demand. In this case, two Norwegian schools with CO_2-sensor based demand controlled displacement ventilation (DCDC-CO_2) and traditional constant air volume (CAV) mixing ventilation were compared [5].

The results showed that DCDV-CO_2 reduced the ventilation air volume by 65–75% in both schools compared to CAV and that after a sampling period of a week it was found that DCDV-CO_2 daytime operation weekdays reduced the total heating energy demand by 21%, the amount of unrecovered heat in the exhaust ventilation air by 54% and the average airflow rate by 50%. Furthermore, it was concluded that fan energy consumption under DCDV-CO_2 daytime operation reduced the fan energy consumption by 87%.

Once the control system is improved, new future parameters like infection risks between environments must be investigated. In this sense, from its origins, the CFD studies were related with ventilation maps, and different applications like operating theatres were developed [6]. This kind of research works concluded the importance of preventing infection in thermally comfortable operating theatres and that reducing airborne infection risk is an important factor in the developments of operating theatres layout and ventilation schemes. Despite this, more investigation based on experimental measuring and its simulation must be developed. These two fundamental points will be described in the next sections.

1.2. Research on Ventilation Methods

To solve most of problems commented in the previous section some typical studies were developed centred on experimental measurements, simulation software and, finally, experimental contrast between field measurements and simulation results. In this section principal research procedures are discussed in accordance with the most recent research works.

1.2.1. Experimental Measuring of Natural Ventilation

Nowadays, new techniques are employed to define the behaviour of a naturally ventilated building. These experimental techniques can be classified into new sampling procedures, wind tunnel and chamber experiments and CFD simulations. All these procedures will be related with the most recent research works in this section.

Despite the fact that the tracer gas is an interesting procedure to define the air changes of an indoor ambience, it needs a particular gas that must be released. In consequence, the sample process needs some anticipation to prevent the quantity of gas that must be released and the sensors working period and maintenance. As a solution to this problem, recent research works showed the feasibility to employ a heat source to trace the ventilation rate in an air inlet/outlet [7]. The idea was to introduce the heat source perpendicular to the direction of the flow opening and measure temperature decay very close to the heat source (0.13 m). In consequence, in these experiments, the 2D temperature field in the flow direction was recorded using an infrared camera. Its results showed that, if adequate heat is applied, it is possible to estimate the ventilation rate with 15% inaccuracy using temperature information from the arc. This new technique let us sample without considering the quantity of released gas and relate this with electronic control systems. In consequence, it appears to be an interesting future methodology to be employed to control indoor ambiences.

To define the real behaviour of a building indoor ambience over different wind conditions, parameters like loss factor must be defined in experimental laboratory studies, like wind tunnel to employ the obtained data in the corresponding Computational Fluid Dynamics software database. For example, to investigate the influences of opening size, thickness ratio,

Reynolds number and door angle on the loss factor a series of wind tunnel experiments were carried out [8]. The experimental data show that this loss factor depends on the internal porosity and door angle and is independent of Reynolds number. Finally, after wind tunnel experiments, the ventilation rates showed a good agreement compared with the measured velocities at the internal opening and, in consequence, this procedure of resistance model was proposed to be employed in more case studies.

Once the database was developed, computational fluid dynamics software were employed to simulate real ambiences. For example, the current methodology to define the behaviour of a natural cross-ventilation with openings at non-symmetrical locations is by test chamber experiments and using advanced computational fluid dynamics techniques [9]. From these studies it was concluded that all turbulence models agree well with the experimental results and, in consequence, this procedure must be employed in industrial ambiences to predict the indoor environments in future buildings.

1.2.2. Simulation of Ventilation Methods

1.2.2.1. Natural Ventilation and Hybrid Ventilation Models

To improve the energy saving in natural ventilation some models were proposed to predict mixing ventilation and upward displacement ventilation. In particular, ref. [10] showed different models of each of these ventilation procedures and a significant energy saving was predicted with mixing ventilation. Despite this, the energy saving from the mixing ventilation can only be achieved if the internal heat gains are significant.

To simulate the transient process of ventilations during the night to purge the heat, recent research works [11] compared two different natural ventilation models: passive displacement and passive mixing during transient conditions. The results showed that, despite the fact that mixing flows were less efficient at purging heat than displacement flows, when a particular zone of the room is considered as a whole, displacement ventilation was more effective than mixing ventilation.

In the last few years, both natural and mechanical ventilation have combined to evolve into a hybrid system. It has been investigated in the IEA Annex 35 "Hybrid Ventilation in New and Retrofitted Buildings" and respective papers [12, 13]. The objective was to develop and adjust ventilation models to reach a comfortable thermal environment, a good indoor air quality with the lower energy consumption. The two numerical models obtained were called HYBCELL1.0 and SPARK and adjusted with experimental results developed in a test classroom with mechanical (fans with or without heat recovery) and natural ventilation (two inlet grilles and an exhaust chimney). The main results were as follows:

1. In winter, hybrid ventilation system has higher energy consumption but there is a significantly lower mean and maximum CO_2 concentrations compared to the mechanical system. There is a better indoor air quality for the hybrid ventilation system with equivalent thermal comfort in terms of indoor air temperature for the three ventilation systems.
2. In spring, hybrid ventilation system has higher energy consumption with better indoor air quality and thermal comfort. However, differences in energy consumption between the three ventilation systems are lower than in winter.

3. In summer, hybrid ventilation energy consumption is lower than reference systems but the CO_2 concentrations are a little bit higher.

As mentioned before, mixing and hybrid ventilation are gaining special interest.

In general, results showed a good agreement between developed models and experimental results and, in consequence, it can be employed to predict the relative performances of ventilation systems. This will be analyzed in the next section.

1.2.2.2. Indoor Ambiences Simulation

In the last years, CFD emerged as an adequate tool to predict indoor ambiences in buildings. In particular, ref. [14] showed a good agreement with wind tunnel and real sampled data. This tool was employed to predict the wind-driven ventilation behaviour of a continuous roof ventilator in a twin-span Mediterranean-type greenhouse. Its results showed that green houses with length less than 50 m have a higher ventilation rate.

Nowadays, more research works about the design of natural ventilation in buildings are performed with computational fluid dynamics (CFD) techniques [15]. In particular, the model based on renormalization group (RNG) theories showed a good agreement with experimental data. Despite the fact that large-eddy simulation (LES) is the most common model it is preferable in wind engineering application and advanced fluid dynamics [16]. In these last cases, LES showed a good agreement with experimental data, in particular, with the predicted airflow patterns and velocities around and within the buildings, and the surface pressure.

Finally, new models that define the expected air changes in indoor ambience based on natural ventilation are developed. These models define the expected air renovation in some cities, considering parameters like wind and thermal buoyancy forces [17]. It was concluded that the hourly wind speed and temperature difference between indoor and outdoor are the most important factors to define the potential of natural driving forces. These kinds of simplified models are expected as a good tool to be employed at the design moment to improve the ventilation system of buildings and reduce their energy consumption.

Once some models are developed, they are used to define the expected parameters like the natural ventilation cooling potential (NVCP) in some important cities to predict parameters of general interest like, for example, the effect of weather over energy consumption and the future climatic change.

Recent researches [18] showed that the NCP depends on the multiple impacts of climate, thermal characteristics of buildings, internal gains, ventilation profiles and regimes. For example, if a climate has a lower monthly mean, external air temperature has higher NVCP. Furthermore, in the very cold zones, day and night cross-ventilation is sufficient for summer cooling in most cases. In the mild zones, full natural ventilation is highly recommended for an office building, and for offices with high internal gain, day and night cross-ventilation has higher efficiency. In hot summer and warmer winter, natural ventilation could not meet the space-cooling requirements, so mechanical cooling or air conditioning is desirable. Another example is the insulation level of external envelope. If external envelope (wall and window) is higher, the NVCP also increases. On the other hand, it was observed that internal heat gain has a significant impact on NVCP.

Nowadays, new cross natural ventilation models [7] improved with human window open behaviours, wind velocity and humidity showed that there are more natural ventilation hours than previous models estimated, especially in cold climate regions. This was related with the fact that window opening percentage should be considered. Furthermore, parameters like wind velocity increases the upper limit comfort temperature and humidity does reversely. Finally, parameters like noise and pollution are proposed to be considered in future research works.

1.3. Personalized Ventilation

A new ventilation procedure is the personalized ventilation (PV). This methodology pretends to supply clean air directly to the breathing zone of occupants and provides the possibility for individual control of the microenvironment at each workstation. Recent research works attempted to define the efficiency of a personal ventilation system were developed [20]. In this sense, ventilation effectiveness was measured on a prototype system for local air conditioning. To do it, measurements of velocity field and of tracer gas concentration were taken to define a parameter based on the relative decrease of tracer gas in the first minute of system operation. That was employed as a methodology to define the efficiency of ventilation systems.

Another advantage of personalized ventilation system is related with the comfort. In this sense, the thermal comfort level in desks with two personalized ventilation systems in slightly warm environments was sampled [21]. Its results showed that the predicted percentage of dissatisfied people reduced from 27.77% without personalized ventilation to 16.1% with personalized ventilation. Despite this, the draught risk and, in general, the thermal discomfort were located in the left human body sections. As a general conclusion, it could be said that the classroom desk design with two personalized ventilation systems guaranteed acceptable thermal comfort conditions and good air quality with acceptable local thermal discomfort conditions.

Similar research works showed that the use of personalized ventilation (PV) system in the same hot and humid climate, in conjunction with a background or ambient air conditioning system not only enhances thermal comfort and IAQ acceptability but can also reduce energy consumption by 15–30% [22]. Furthermore, it was found [23] that the use of desk-based PV in conjunction with mixing background ventilation may reduce the energy consumption up to 51% compared to mixing ventilation alone with some of these strategies:

a) Reducing the airflow rate due to the higher ventilation effectiveness of PV.
b) Increasing the maximum allowed room air temperature due to PV capacity to control the microclimate.
c) Supplying the outdoor air only when the occupant is at the desk.

A special kind of personalized ventilation system is the ceiling-mounted PV system with supply air terminal devices (ATD). This system, installed on the ceiling, does not require ducting for supply of outdoor air to each workplace and thus improves indoor aesthetics and provides a considerably higher level of flexibility in the furniture layout. For this kind of ventilation procedure some conclusions were obtained:

- The use of PV with ceiling-mounted ATD in conjunction with mixing ventilation at an elevated room temperature of 26ºC will decrease the total energy consumption in comparison with mixing ventilation aiming at a room air temperature of 23.5ºC.
- The use of PV with ceiling-mounted ATD in conjunction with mixing ventilation at a room air temperature of 26ºC will lead to use of substantially less energy in comparison with energy used with mixing ventilation at 26ºC and 28ºC and desk fans for providing thermal comfort at each workplace.
- At an elevated room temperature of up to 28ºC the use of desk fans for providing convection cooling to each occupant in rooms with mixing ventilation may lead to increase energy consumption in comparison with the energy used to keep comfortable room air temperature by mixing ventilation alone.

Now that the main advantages of personalized ventilation systems respect the displacement ventilation were shown some new disadvantages like the effect of walking persons were investigated [24]. In particular, this disadvantage is of special interest over "ductless" PV. In this situation the PV sucks, transports and supplies the clean and cool air distributed over the floor area by the DV, to the breathing zone of the occupant. In this way, a clean and cool air can be better utilized and the flexibility of desk layout is not affected [24]. The results showed that the performance of the "ductless" personalized ventilation was better than the performance of the displacement ventilation used alone. The walking persons diminished the vertical temperature gradient in the vicinity of the workplaces and increased the inhaled air temperature. Under all the conditions (with and without walking) the temperature of the air inhaled by the manikins under the "ductless" personalized ventilation was lower than with displacement ventilation alone. Thus in practice the "ductless" personalized ventilation will be superior to displacement ventilation alone with regard to the perceived air quality. On the other hand, at elevated room air temperatures the "ductless" personalized ventilation in conjunction with displacement ventilation will provide occupants with a better thermal comfort than displacement ventilation alone because of the increased local body cooling by the personalized flow.

1.4. Natural Ventilation and Energy Saving Methods

Natural ventilation was evaluated in the last years as the low energy methodology to renovate building indoor environments. In this sense, energy saving is related with the improvement of building design, based on passive methods [25-28], and in the improvement of the HVAC system.

An example of passive methods for reducing the energy saving in ventilation is solar-induced ventilation [29] which is proposed as an alternative in enhancing the stack ventilation when it experiments weather conditions that make it inefficient, due to small temperature difference between the inside and outside of buildings. To summarize, the effectiveness of this solution is defined through the air temperature difference between the air inside the stack and the ambient air.

From this research work it was concluded that this procedure is able to create air temperature difference of more than the usual air temperature difference attained in naturally ventilated buildings. Despite this, parameters like wind have significant effect on the induced air velocity by the proposed strategy and it can be beneficial or detrimental depending on the position of inlet and outlet.

After the HVAC system is installed and works in a particular building only some maintenance techniques like the improvement of the control system can reduce the energy consumption. In this sense, recent techniques are being developed to improve the natural ventilation and its associated energy consumption. For example, improvement in demand-controlled ventilation (DCV) was analyzed by ref. [30]. In particular, two approaches, CO_2-DCV and RFID-DCV, are proposed to be applied in DCV simulations and compared with ASHRAE and the standards of different countries.

It must be clarified that RFID is the radio frequency identification (RFID) technology. By attaching an RFID tag on a personal identification card, the exact number of occupants in each room can be traced. In consequence, RFID-DCV method adjusts the amount of ventilation air based on the real-time number of occupants.

When DCV was simulated in accordance with the ASHRAE Standard 62.1-2007, ventilation air to the space was defined as the sum of two ventilation components: people and building components. The first term, the people component, is the product of the real time-number of people and the minimum ventilation rate per person. The number of occupants is estimated from the change of indoor CO_2 concentration in the CO_2-DCV case and precisely measured in the RFID-DCV case. The second term, building component, is acquired by multiplying the ventilation rate per unit floor area by the total floor area of the given space.

On the other hand, when we try to simulate the DCV, in accordance with Korean ventilation standard for multipurpose facility, the ventilation rate per person required to control contaminants emitted from both the people and the building must be defined together; that is, people and building components are not defined separately as ASHRAE Standard 62.1 do.

Simulation results showed that the current Korean standard might cause unstable DCV behaviour, especially under CO_2-DCV criteria due to the fact that the air changes are related with internal sources and not only the number of occupants. In consequence, the current occupant-based ventilation standard was modified as a form of ASHRAE Standard 62.1-2007 showing good applicability to various DCV approaches and a reduction in fan energy consumption.

As a general conclusion, we can say that to improve the DCV control system a separate consideration of each air change criterion must be developed. Despite this, new presence detection sensors can help us to find an approach to the real number of occupants of an indoor ambience in each moment and, in consequence, reduce the energy consumption within the minimum air changes required for each indoor ambience.

REFERENCES

[1] Niachou, K.; Hassid, S.; Santamouris, M.; Livada, I. *Building and Environment.* (2008). Experimental performance investigation of natural, mechanical and hybrid ventilation in urban environment. 43, 1373-1382.

[2] Tung, Y.; Shih, Y.; Hu, S.; Chang, Y. Building and Environment. (2010). Experimental performance investigation of ventilation schemes in a private bathroom. 45, 243-251.

[3] Bekö, G.; Lund, T.; Nors, F.; Toftum, J.; Clausen, G. Building and Environment. (2010). Ventilation rates in the bedrooms of 500 Danish children. 45, 2289-295.

[4] Caputo, A.C.; Pelagagge, P.M. (2009). Applied Thermal Engineering. Upgrading mixed ventilation system in industrial conditioning. 29, 3204-3211.

[5] Wachenfeldt, B. J.; Mysen, M.; Schild, P. G. (2007). *Energy and Buildings*. Air flow rates and energy saving potential in schools with demand-controlled displacement ventilation. 39, 1073-1079.

[6] Chow, T.T.; Yang, X.Y. (2004). *Journal of Hospital Infection*. Ventilation performance in operating theatres against airborne infection: review of research activities and practical guidance. 56, 85-92.

[7] Özcan, S.E.; Vranken, E.; Berckmans, D. (2009). *Building and Environment*. Measuring ventilation rate through naturally ventilated air openings by introducing heat flux. 44, 27-33.

[8] Chu, C.; Wang, Y.. (2010). *Building and Environment*. The loss factors of building openings for wind-driven ventilation. 45, 2273-2279.

[9] Stavrakakis, G.M.; Koukou, M.K.; Vrachopoulos, M.Gr.; Markatos, N.C. (2008). Energy and Buildings. Natural cross-ventilation in buildings: Building-scale experiments, numerical simulation and thermal comfort evaluation. 40, 1666-1681.

[10] Woods, A. W.; Fitzgerald, S.; Livermore, S. (2009). *Energy and Buildings*. A comparison of winter pre-heating requirements for natural displacement and natural mixing ventilation. 41, 1306-1312.

[11] Coffey, C.J.; Hunt, G.R. (2007). *Building and Environment*. Ventilation effectiveness measures based on heat removal: part 2. Application to natural ventilation flows. 42, 2249-2262.

[12] El Mankibi, M.; Cron, F.; Michel, P.; Inard, X.C. (2006). *Solar Energy*. Prediction of hybrid ventilation performance using two simulation tools. 80, 908-926.

[13] Gan, G. (2010). *Energy and Buildings*. Simulation of buoyancy-driven natural ventilation of buildings-Impact of computational domain. 42, 1290-1300.

[14] Mistriotis, A.; Bot, G.P.A.; Picuno, P.; Scarascia-Mugnozza, G. (1997). Agricultural and forest meteorology. Analysis of the efficiency of greenhouse ventilation using computational fluid dynamics. 85, 217-228.

[15] Evola, G.; Popov, V. (2006). *Energy and Buildings*. Computational analysis of wind driven natural ventilation in buildings. 38, 491-501.

[16] Jian, Y.; Alexander, D.; Jenkins, H.; Arthur, R.; Chen, Q.(2003). *Journal of Wind Engineering and Industrial Aerodynamics*. Natural ventilation in buildings: measurement in a wind tunnel and numerical simulation with large-eddy simulation. 91, 331-353.

[17] Yang, L.; Zhang, G.; Li, Y.; Chen, Y. (2005). *Building and Environment*. Investigating potential of natural driving forces for ventilation in four major cities in China. 40, 738-746.

[18] Yao, R.; Li, B.; Short, A. (2009). *Renewable Energy*. Assessing the natural ventilation cooling potential of office buildings in different climate zones in china. 34, 2697-2705.

[19] Yin, W.; Zhang, G.; Yang, W.; Wang, X. (2010). *Building and Environment.* Natural ventilation potential model considering solution multiplicity, window opening percentage, air velocity and humidity in china. 45, 338-344.

[20] Muhic, S.; Butala, V. (2006). *Energy and Buildings.* Effectiveness of personal ventilation system using relative decrease of tracer gas in the first minute parameter. 38, 534-542.

[21] Conceiçao, E.Z.E.; Lúcio, Mª. M. J.R.; Rosa, S.P.; Custódio, A.L.V.; Andrade, R. L.; Meira, M.J.P.A. (2010). *Building and Environment.* Evaluation of comfort level in desk equipped with two personalized ventilation systems in slightly warm environments. 45, 601-609.

[22] Yang, B.; Sekhar, C.; Melikov, A. K. (2010). *Energy and Buildings.* Ceiling mounted personalized ventilation system in hot and humid climate- An energy analysis. doi: 10.1016/j.enbuild.2010.07.022.

[23] Schiavon, S.; Melikov, A. K.; Sekhar, C. (2010). *Energy and Buildings.* Energy analysis of the personalized ventilation system in hot and humid climates. 42, 699-707.

[24] Halvonová, B.; Melikov, A. K. (2010). *Building and Environment.* Performance of "ductless" personalized ventilation in conjunction with displacement ventilation: Impact of disturbances due to walking persons. 45, 427-436.

[25] Orosa, J.A.; Oliveira, A.C. (2010). *Energy and Buildings.* Implementation of a method in EN ISO 13790 for calculating the utilisation factor taking into account different permeability levels of internal coverings. 42 (5), 598-604.

[26] Orosa, J.A.; Oliveira, A.C.(2009). *Renewable Energy.* Hourly indoor thermal comfort and air quality acceptance with passive climate control methods. 34 (12), 2735-2742.

[27] Orosa, J.A.; Oliveira, A.C. (2009). *Energy and Buildings.* Energy saving with passive climate control methods in Spanish office buildings. 41 (8), 823-828.

[28] Orosa, J.A.; Oliveira, A.C.(2010). *Energy and Buildings.* Implementation of a method in EN ISO 13790 for calculating the utilisation factor taking into account different permeability levels of internal coverings. 42 (5), 598-604.

[29] Yusoff, W.F.M.; Salleh, E.; Adam, N. M.; Sapian, A.R.; Sulaiman, M.Y. (2010). *Building and Environment.* Enhancement of stack ventilation in hot and humid climate using a combination of roof solar collector and vertical stack. 45, 2296-2308.

[30] Jae-Weon, J.; Anseop, C.; Sang-Tae, N. (2010). Simulation Modelling Practice and Theory. Improvement in demand-controlled ventilation on multi-purposed facilities under an occupant based ventilation standard. 18, 51-62.

Index

A

B

C

D

E

F

J

K

L

M

Q

R

S

Y